GILL MCKAY

STUCK

BRAIN SMART INSIGHTS
FOR COACHES

RETHINK PRESS

First published in Great Britain in 2019
by Rethink Press (www.rethinkpress.com)

Cover image © Shutterstock/Vladgrin
Printed and bound by CPI Group (UK) Ltd, Croydon, CR0 4YY

Praise

'As a coach of eighteen years I am grateful to this book for providing the essential gravitas needed to enable coaches and their clients to understand that real, sustainable change can occur.

With this understanding of how the brain operates, how to manage and work with our mindsets shows us that all potential is possible. Useful, wise and a much needed add-on to a professional coach's tool kit and offer.'

Carole Ann Rice

Coach and Happy Monday columnist for the *Daily Express*

'Most people have had an experience of being stuck in their life or work, not knowing which way to turn or whether there is a way forward. Reading Gill's book will help you to understand those stuck times and the underlying neurological mechanisms that make those periods challenging. More importantly you will gain an insight into how you can make choices that will lay down new neural connections for getting back on track and opening up new possibilities. *STUCK: Brain smart insights for coaches* is full of case studies, stories and examples of strategies to move you forward and is a valuable resource for you to use with clients if you are a coach, with your team if you are a leader, or for yourself if you are curious to understand how you can get your brain to work with you and not against you.'

Luca Vanni
Vice President HR & Organisational
Effectiveness, EMEA, NEC Europe

'Gill has achieved a unique bridge in what are, in many ways, two conflicting areas in the role of being a coach. Whilst it may be a slight truism, coaches tend to be "people" people who love finding patterns, holes and triggers. We don't tend to enjoy the small bits... the details. Many coaches are extremely intuitive and a bit frightened of the science: that's me by the way! But Gill's book was an easy read in a very complex area. I have read many books in the neuroscience area over the last twenty years, most written by neuroscientists and I've only really been able to take away top level conclusions. But STUCK has broken down the science into understandable blocks and brought it to life through case studies. In addition, Gill's warm and caring personality shines through, making the read not only rewarding from a learning point of view, but also a pleasure.'

Mark Newey
Therapeutic Coach and author of *The Naked I: Authenticity: Be you be happy* and contributing author to *Boys Do Cry. By 12 men who did. Woke up. And redefined what it means to be a man*

'The challenges presented by the acceleration and scale of change in our work and personal lives compel us all to operate differently. For many, coping with change is reactive. For others it's a choice about how to think in a different way about these challenges. *STUCK* helps us with thinking differently, in an accessible and engaging style, through recent research into the exciting hot topic of neuroscience. Using the power of stories and case studies from her coaching work, Gill explains the neuroscientific underpinnings common to business and personal challenges. Whether you're a coach or looking for coaching ideas, *STUCK* provides practical, brain-based tools and insights to meet these challenges. In order to truly operate at our best, we all need to understand how the brain enables change and transformation and why certain actions help us to progress. This book is a compelling read for both business leaders and coaches choosing to think differently.'

Charlotte Campbell

IT sector business leader in sales enablement
and transformation at StrataCom, Cisco Systems,
EMC, Salesforce.com and Dell Technologies

'The world of coaching is maturing. In turn coaches and their clients are demanding more. *STUCK* is a highly useful addition to the repertoire of knowledge and understanding around human behaviour. Gill McKay cleverly uses a 'left brain/right brain' approach to the whole book, using science and narrative to bring alive what can otherwise be impenetrable and out of reach. Using plain language and metaphor, a whole new dimension of neurology is opened up to us. It reveals why we get stuck in habitual ways, and how we might break out of these using evidence-based understanding of the brain.

Gill has done the hard work of locating research and pertinent examples that will allow the reader to swiftly enrich their work on themselves or with others. Too many of us can be daunted by science. But by describing it in bite-size chunks and weaving stories throughout, a whole new layer and language is made meaningful and exciting. After twenty years working as a coach, reading *STUCK* has given me new ideas and some better explanations for behaviours and patterns I observe. I trust it will do the same for many readers and give us another helpful measure of authority and confidence to go further and deeper into this fascinating world of behaviour change.'

Alasdair Cant
Director of Cambridge Training

This book is dedicated to my parents who sadly are no longer around to see that I finally did it.

Contents

Introduction

It is a privilege to be a coach, to be part of a person's journey to improvement, to enlightenment, to a better path. Knowing that in some small way, a conversation I have been part of has moved my client on, helped them increase awareness, gain insight, and discover new possibilities is rewarding beyond measure. In my career of more than thirty years, I have worked predominantly in the learning and development field, both employed in the corporate world and as an independent consultant, now as co-founder of MyBrain International. I work with individuals and groups, as a coach, trainer and facilitator – and at the core of everything I do are the themes of awareness, development, change, strengths, flow and motivation, enabling people to be the best they want to be.

Gaining formal coaching qualifications and training in neuro-linguistic programming opened up new personal discoveries and interests in the early 2000s. I am fascinated with neuroscience, a subject which is now paving the way for deepening the scientific understanding of human behaviour and psychology. The last twenty years have shone an ever-brightening light on this emerging area; no longer do neuroscientists rely solely on autopsies, patients with extreme brain conditions or inferences from laboratory rodent experiments to understand the workings of the brain. Scanning technology improvements have fulfilled a dream of providing images of activity inside the brains of healthy and functioning human beings. Recognising that scanning procedures are not without limitations, the saying 'a picture is worth a thousand words' rings true as visual images are helping to deepen our understanding of the physiology behind human behaviour and further scientific discussion. Moreover, the insights gained from neuroscience are no longer just relevant in the medical and mental health spheres, but increasingly can be applied in business. The wider exploration in the non-clinical context is providing evidence that is appropriate in multiple areas of people development, performance, effectiveness, individual differences, processing styles, motivation, and happiness to name a few. This has opened up exciting possibilities for anyone interested in the development of self and others.

I wrote this book to share some coaching stories and how insights from neuroscience have helped my clients. Each client has graciously given permission for me to describe their story in the spirit of learning – names and some context have been changed to honour confidentiality. The issues presented are varied but there are three patterns that are common across each. Firstly, in all cases the client was stuck. Wanting to be better and make some changes is the heartbeat of my coaching conversations. The coaching sessions created space for clients to pause, unpick their challenges, recognise triggers, acknowledge feelings, voice concerns and set a new path. Secondly, whether their feeling of being stuck manifested as a crossroads of possibilities or a downward-spiralling, aimless dead end, each client was experiencing wavering confidence. They weren't sure what their next step was, were hesitant, doubtful, sometimes even lost. Finally, each client possessed a curiosity about what made them tick, what was happening inside their brains that they could understand and use to help them progress. Including the dimension of neuroscience to the coaching was a useful addition in all cases, broadening out the conversation and possibilities for the future. Becoming aware of how their brains worked helped all my clients to make progress and become unstuck.

This book is for both coaches and clients. If you are interested in self-development and identify with some of the client stories, then the brain insights will add an additional perspective. As a coach, you will recognise familiar client scenarios and see how the layer of neuroscientific knowledge can deepen a coaching conversation. True coaching is about client exploration, unlocking their resourcefulness, enabling them to make discoveries and identify possibilities. In respecting these core principles around coaching as an enabling conversation, I thought long and hard about introducing an element of 'teach' into my sessions. For many clients, some of the neuroscience research is new information, so the coaching focus moves at this point. Across each unique client issue and context, I am delighted that the coaching has been enriched by adding this additional layer, broadening out the discussion, opening up new avenues and exposing new insights. Discussing the brain also equipped me to ask questions of my clients at a deeper level, eliciting responses I don't believe would otherwise have been teased out. I hope you find that as well in your future coaching work and that this book helps in adding that layer to your skills.

Each chapter starts with a client story – why they were stuck, their background, context and viewpoint. I then describe the neuroscientific evidence in the realm of the presenting issue and the work I undertook with the client, explaining how I introduced the neuroscientific element and how it broadened the coaching discussion. There are six chapters examining six different presenting issues: habit change, stress, negative thinking, loneliness, decision-making and motivation. In each, I highlight discrete neuroscientific research pertaining to specific functions of the brain known to be impacted by that presenting issue. It is essential to remember that this is a simplification – regions of the brain work in conjunction with each other, brain circuits interact, most areas are dependent on others and have multiple functions. My purpose is to shine a light on critical, up-to-date research and its application and inter-relations with different brain regions is highlighted throughout. Chapter 7 concludes by shifting the focus to us as coaches showing how integrating neuroscience into our coaching conversations and becoming consciously aware of brain functioning can add to both our client's enjoyment of the interaction, and our effectiveness, helping us to show our clients the very best version of ourselves.

This book comes with a warning. It is not a 'one-stop instruction guide to neuroscience and coaching'. You will not learn everything you need to know about neuroscience and embedding the subject into your coaching sessions. How can you when the subject is so new and evolving? Nor is it packed with generic neuro-tips, tools and techniques. As a coach, you will be skilled and adept at engaging with your clients, yet a prescriptive, sexy tool kit implies a formulaic, one size fits all approach – the antithesis of what coaching is all about. Rather, my intention with the book is to whet your appetite to keep learning about our wonderful brains through real stories from my clients and the work we have done together embracing neuroscientific findings. All research and quotes are referenced, many of which I have read for my own learning about the amazing subject of neuroscience as well as for high-lighting specific evidence for each presenting issue. Please have a look at our website www.mybrain.co.uk for articles and many detailed book reviews, which we frequently add to.

Every day we are all bombarded with neuro-nonsense in newspapers, blogs, social media and magazines where a research conclusion has been distorted and misinterpreted as the truth. My hope is to bridge the gap between that neuro-nonsense and the empirical academic research studies and offer evidence that is accessible, useful, current and interesting. Every day, neuroscientists, psychologists and behavioural scientists worldwide are on the point of new discoveries about the human brain. These new insights can help us all become more aware and develop further. That is truly exciting.

PART 1

A HELICOPTER VIEW OF THE HUMAN BRAIN

During the last 100 years humans have developed computers, mapped all of the genes in the human genome, sent satellites to the outer reaches of the galaxy and put a man on the moon, yet we are probably many generations away from truly understanding how the human brain works. With its billions of neurons and trillions of connections, the human brain is by far the most complex thing known to man, yet what we are just discovering now enables us to form a basic understanding of how our brains make us the people we are and our knowledge is constantly developing.

The descriptions found in this first section are provided at a high 'helicopter' level to give you an overview of the major brain areas. You can see them summarised in the illustration overleaf and all the brain regions highlighted in the extensive research throughout this book are included.

THE BRAIN IS YOUR CONTROL CENTRE

Hindbrain

Helps to regulate automatic functions,
coordinate movement, maintain balance
and equilibrium and relay sensory information.

Forebrain

Processes sensory information, helps
with thinking, planning, reasoning, language
processing, problem solving and regulates
automatic, endocrine and motor functions.

Midbrain

Plays a role in physical motor actions and
helps to process auditory and visual information.

It receives information through your five senses assembling messages so they have meaning for you.

Forebrain

The forebrain or prosencephalon makes you uniquely human and is your largest brain area consisting of the diencephalon and the telencephalon.
The diencephalon includes the thalamus, hypothalamus, pituitary gland and the pineal gland.
The telencephalon includes the cerebral cortex, corpus callosum and limbic system.
The cerebral cortex includes the frontal, occipital, parietal and temporal lobes.
The frontal lobes include the prefrontal cortex which further breaks down into the inferior, dorsolateral, ventromedial and orbitofrontal cortices.
The limbic system includes the hippocampus, amygdala, cingulate and the insula.*

Midbrain

The midbrain or mesencephalon is the uppermost part of the brainstem.
It consists of the tectum and a number of basal ganglia structures - the substantia nigra, striatum, the nucleus accumbens, caudate nucleus, putamen and ventral tegmental area.**

Hindbrain

The hindbrain or rhomebencephalon is the most primitive part of your brain, located at the base of your skull. It is made up of three main parts - the medulla oblongata at the bottom of the brainstem, the pons just above it and the cerebellum at the back.

Sight

Smell

Taste

Touch

Hearing

*There is some disagreement between neuroscientists on the grouping of some of these brain structures. See text for details.
**The basal ganglia are large subcortical structures comprising several interconnected nuclei in the forebrain and midbrain but are described in the midbrain section. See text for details.

How to use Part 1

The neuroscientific descriptions and explanations in Part 1 are offered at a high level to provide an overview of the major brain areas and enough information to give some depth and substance to your understanding. This section of the book can be used as a reference guide as you discover research findings relating to client presenting issues. Or of course you can read it through as a way to build knowledge before looking at the context of different presenting issues in Chapters 1–6 in Part 2. Chapter 7 summarises this brain knowledge by showing the relevance and value of integrating conversations about specific brain regions and research into your coaching work.

Why Do You Have a Brain?

Have you ever considered this question? The answer may be a surprise – it's not to think or feel or even to run your body. The fundamental reason you have a brain is because you move and need to interact dynamically with the world around you. To survive, as a human you need to be able to move around your environment in a meaningful way. Anthropologists regard the Australopithecus as being your oldest ancestor – a creature that evolved in eastern Africa around 4 million years ago. They were the first hominids to show the presence of the SRGAP2 gene, which enables neurons to develop increased length and functionality.[1]

During the following 4 million years human brains have roughly trebled in size as their functionality increased. For the vast majority of that time your ancestors' only priorities were food, shelter, warmth, safety, and continuation of the species. They lived in a natural environment in woods or in caves, which changed at the pace of the seasons. They relied on instincts and some memories so they could repeat successful patterns of their existence from season to season. Since then, your brain has trebled in size and developed a frontal cortex for cognitive thinking functions through evolution, however the parts of your brain that provide your instincts and automatic functions that your ancestors relied on still account for around 80% of your neurons.

It's all connected

I have focused on specific brain regions so you can build an understanding of some of their primary functions and responsibilities. Although a human brain looks like a single mass of jelly, it is in fact made up of many different parts, with each part responsible for different functions and for controlling different aspects of our lives. In broad terms we now know what these different parts of the brain are responsible for, but we also know that if one part becomes damaged, other parts of the brain will take over some of its work and that some functions of the brain require a multitude of individual parts of the brain to interact and operate together.

It is important to recognise that each region has many connections with other regions and different regions can be part of multiple brain circuits. Simply put, brain regions need each other to work at their best. Think about the financial markets. Analysts and dealers all over the world are buying and selling 24/7, currencies are fluctuating, companies are announcing results, tech coins are launching, forex robots and algorithms are at work. Stock markets all over the world are volatile, are dependent on multiple factors and interact dynamically. It doesn't take much for a trade to crash and tumbling results spread like a virus. It's the same with the brain, if just one region experiences a blip, or a neural circuit is operating sub optimally, there will be knock-on effects throughout. It's all connected.

The wonders of technology

The development of different scanning technologies in recent years has provided fascinating new insights into the workings of the brain, but there is often disagreement in the scientific community over the interpretation of the data produced by these scans. Although knowledge is advancing all the time, it is unlikely that a complete understanding of the brain will be achieved within your lifetime, if ever. A high-level summary of scanning technology referred to in this book is provided later in this section.

The basic building blocks of the brain

A neuron is a specialised type of cell found predominantly in the brain and the central nervous system. A neuron has two special features – firstly it is electrically excitable and secondly, it connects using synapses. A typical neuron is divided into three parts: the cell body (or soma), dendrites, and axon. Dendrites look like branches of a tree and contain the receptors that receive the chemical messages, known as neurotransmitters, from other cells. The number of dendrites per neuron can vary from just one or two up to over a thousand. Neuroscientists suggest that the number of dendrites provides an indication as to the functional complexity of the neuron.

Axons are fine strands that extend from the soma along which electrical messages are passed. They can be more than a metre in length where they pass down the spinal cord, which itself consists of millions of axons. Axons are protected by a myelin sheath, which helps the transmission of the electrical current and insulates them in a similar way to plastic surrounding electrical cabling. At the end of the axon are axon terminals which convert the electrical messages from the soma into chemical neurotransmitters that are then released into the synaptic gap to communicate with other neurons.

Synapses are the gaps between the neurons. To communicate, the electrical signal from one neuron must jump across these gaps. Your brain achieves this by turning the electrical signal into a chemical message at the axon terminal. These chemical messages can only be received by a neuron with a matching receptor in much the same way as a child's hammer and shapes toy works – the pegs will only fit in the holes of the same shape. Unlike the toy, the synaptic network is amazingly complex. Each neuron can establish connections to thousands of other cells in the body with the number of connections reaching a peak by the time you are around three years old.

The development of the brain

No two brains are the same. Just as your fingerprints and retinas are unique, so is your brain. It all begins in the womb when, around seven weeks after conception, your first neurons and synapses begin to develop in your spinal cord. During the next few months your brain will develop to eventually reach around 86 billion neurons.[2] About sixty days before you are born, each neuron begins to establish connections with other cells in the body. In some cases, these connections are electrical and in other cases chemical. The number of connections each neuron will establish will vary from a few thousand up to 100,000. By the time you are three, you will have an average of around 10,000 synaptic connections for each neuron.

To put that into perspective, suppose that the Amazon Rain Forest, which consists of roughly 2.4 million square miles of forest, contains around 86 billion trees. Each tree will have around 10,000 leaves. Your brain is therefore akin to all the leaves in the Amazon Rain Forest making it by far the most complex organ in your body. The number of synaptic connections in the brain peaks in early childhood, around the age of three and as you move from being a baby to becoming a toddler, the process of synaptic pruning begins. The synaptic connections you use on a regular basis become stronger, and those that you do not use get weaker, or even disappear altogether.

By the time you are sixteen you will have lost around half of your synaptic connections and the basic foundation of who you are, your character and your values will have been established. Psychologists call this the imprint period. Although your personality may be relatively well established, this does not mean that it does not change thereafter. Your brain's ability to continue to reorganise the neural pathways in the light of experience is referred to as neuroplasticity.

The process of synaptic pruning continues into early adulthood, not surprisingly as the prefrontal cortex does not reach full maturity until around the mid-twenties. As you learn and change the way you approach things, you also develop new synaptic connections. Although the overall number of synaptic connections tends to decline throughout your life, the nature of the connections within your brain is constantly changing. New connections are formed on an ongoing basis and some connections are lost as part of the brain's normal process of forgetting. The synaptic network of connections is constantly rewiring itself based on your thoughts and experiences.

More recently some neuroscientists have begun to describe the development of the brain as analogous to the development of muscles, as it strengthens with use or, more specifically, the areas that are used most strengthen relative to those that are used less.

As evidence of this, a University College London study of the brains of London taxi drivers in 2000 found that they had larger posterior hippocampi than most people, which the researchers attributed to the extensive amount of navigational memory required for their work.[3] This has led researchers to conclude that the brain develops through the combined effects of both nature (those aspects you inherit from your parents) and nurture (those that result from your environment and experiences). This is due in large part to neuroplasticity which is explained in detail in Chapter 1.

I often use the analogy of a road network rather than a muscle as it helps to explain that the connections within the brain get stronger rather than the brain itself. Imagine that all roads in a newly developed area start off the same size (same number of lanes, same width etc) and that they are made wider or narrower depending on usage. If they are not used at all then they are dug up and houses built in their place. Some roads therefore turn into fifteen-lane super-highways while others become minor back streets.

This is what happens in synaptic pruning. The connections that are used regularly grow, develop and become preferred connections while others wither and disappear.

Grouping the brain into forebrain, midbrain and hindbrain

The brain is a complex organ that acts as the control centre of the body, receiving, sending, processing, and directing sensory information. It receives this information through your five senses – sight, smell, touch, taste, and hearing – and assembles the messages in a way that has meaning for you, storing that information in your memory. The brain controls your thoughts, memory and speech, body movement, and the function of many organs within your body. It is split into left and right hemispheres by a thick band of fibres called the corpus callosum. There are three major divisions of the brain – forebrain, midbrain and hindbrain – with each contributing different roles. These divisions and many of their subdivisions that have been highlighted in studies in this book are discussed below.

The central nervous system is composed of the brain and spinal cord. The peripheral nervous system is composed of spinal nerves that branch from the spinal cord and cranial nerves that branch from the brain.

Forebrain

Most of your uniquely human capabilities come from the forebrain. The expansion of the cerebral cortex, particularly the prefrontal area, is a major evolutionary event in the development of human thinking capability.

The forebrain is the largest brain area consisting of the diencephalon and the telencephalon. These in turn consist of other sub regions, itemised below.

- The diencephalon includes the thalamus, hypothalamus, pituitary gland and the pineal gland.

- The telencephalon includes the cerebral cortex, corpus callosum and limbic system.

- The cerebral cortex includes the frontal, occipital, parietal and temporal lobes.

- The limbic system includes the hippocampus, amygdala, cingulate and the insula.

- The frontal lobes include the prefrontal cortex which further breaks down into the inferior, dorsolateral, ventromedial and orbitofrontal cortices.

There is some disagreement between neuroscientists on the grouping of some of these brain structures. Some think for example that the insula should be considered as a fifth lobe of the cerebral cortex, others part of the temporal lobe within the cerebral cortex and often it is grouped within the limbic structures. As the purpose of this chapter is to show the unique features and purpose of specific brain regions, it doesn't matter where it is grouped for explanation purposes, so I have clustered it with other limbic brain structures.

The diencephalon regions

The diencephalon is a small part of the brain tucked under and between the two cerebral hemispheres, just above the start of the midbrain's brain stem. It is relatively small but has many crucial roles within the central nervous system and is essential for healthy brain and body functioning. It relays sensory information and connects components of the endocrine system with the nervous system. The diencephalon is a key regulator of autonomic, endocrine, and motor functions, also playing a major role in sensory perception. Key components are the thalamus, hypothalamus, pituitary gland and the pineal gland.

Thalamus

The thalamus is a large symmetrical structure that makes up most of the mass of the diencephalon. It is often described as a relay station or gatekeeper because almost all sensory information (apart from smell) passes through the thalamus to reach its destination. It connects areas of the cerebral cortex that are involved in sensory perception and movement with other parts of the brain and spinal cord. It's an important part of cortical processing, also playing a role in the control of sleep and wake cycles.

Hypothalamus

This endocrine structure secretes hormones that act on the pituitary gland to regulate biological processes such as metabolism and growth. It is a small area of the brain located directly above the brainstem and acts as the mothership for homeostasis, respiration, blood pressure and body temperature regulation. It co-ordinates the automatic nervous system and controls the sympathetic and parasympathetic responses. It is also considered a part of the limbic system influencing various emotional responses through its connection with the pituitary gland, skeletal muscular system, and autonomic nervous system.

Pituitary gland

The pituitary hangs from the hypothalamus and is a small pea-sized gland. It is often called the master gland as it controls other hormone glands in your body, such as the thyroid, adrenals, ovaries and testicles.

Pineal gland

This is a small endocrine gland which produces the hormone melatonin, important for the regulation of sleep-wake cycles, seasonal functions and sexual development. It converts nerve signals from the sympathetic nervous system into hormone signals, linking the nervous and endocrine systems.

Cerebral cortex

It includes the cerebrum, which counts for around two-thirds of the brain's total weight, consisting of the cerebral hemispheres covering most other brain structures. The layer of grey matter on the outer surface of the cerebrum is called the cerebral cortex. Cortex is the Latin word for bark and, like a tree, the cerebral cortex is the outermost layer of the brain, containing about 10% of all neurons. It forms extensive connections with subcortical areas, is involved in an enormous amount of brain functions and is essential to their healthy operation. It is sometimes simplified as being made up of three types of areas: the sensory, motor, and association areas.

The surface of the cerebral cortex is made from crinkled layers of folded tissue inside the skull. The folds form bulges called gyri and valleys called sulci that create indentations in the brain in humans and larger mammals. The folding significantly increases its surface area making room for more neurons. If the cerebral cortex of your brain was spread out, it would cover 500 square centimetres. Its thickness varies from 1.5mm in the primary sensory areas to 4.5mm in the motor and association areas.

The cerebrum is divided into two halves: the right and left hemispheres. Each hemisphere controls the opposite side of the body. They are joined by a bundle of 200 million nerve fibres called the corpus callosum that transmits messages from one side to the other. It is the largest pathway in the brain, stretching across the midline. It is c-shaped and is about 10cm long, making up the largest collection of white matter tissue found in the brain.

The cerebral hemispheres have distinct fissures, which divide the brain into lobes. Each hemisphere has four lobes, the frontal, temporal, parietal, and occipital. Each lobe is further divided, into areas that serve specific functions. Please remember when reading any of the following detail that each lobe of the brain does not function alone. There are complex relationships between the lobes of the brain and their sub regions as well as between the right and left hemispheres.

Frontal lobes

The frontal lobes contain the prefrontal cortex, premotor area, and motor area of the brain. These lobes help with voluntary, controlled muscle movement, memory, thinking, decision-making, purposefulness, planning and intentional thinking. They are thought to be responsible for personality, conscience and inhibitions. Neuroscientist Elkhonon Goldberg in his book, *The New Executive Brain: Frontal Lobes in a Complex World*, tells of his mentor, the Russian psychologist Alexandr Luria, referring to the frontal lobes as 'the organ of civilisation'.[4] This 'civilised' brain region provides a wide range of highly evolved functions which will be explored in more detail in this chapter.

Occipital lobes

Lying at the back of the head, the occipital lobes are responsible for receiving and processing visual information from the retina. They interpret vision, colour, light and associated movement. Around 30% of the cortex is used for visual functioning.

Temporal lobes

The temporal lobes are located just above the ears and play an important role in organising sensory input, auditory perception, language and speech production, as well as memory association and formation. The auditory cortex is located in the upper part of the temporal lobe and the inferior and posterior temporal lobe areas are finely tuned for understanding and recognising complex visual objects. An area of the medial temporal lobe called the fusiform face area helps you to recognise faces.

Parietal lobes

These lobes lie at the top of the head in front of the occipital lobes receiving and integrating sensory information. They contain the somatosensory cortex, which processes touch sensations. The parietal lobes are involved in your understanding of numbers and their relations, enabling us to look at a small number of objects (around five) and know how many there are without counting them. Spatial sense, navigation and visual perception are additional roles.

Frontal lobe subdivision – the prefrontal cortex

Neuroscientist Elkhonon Goldberg, whose seminal book I mentioned above, eloquently describes the arrival of the cerebral cortex (in evolutionary terms) as radically changing the balance of power in the brain. The subcortical regions are now subordinated into support roles in the shadow of the new level of neural organisation. He says:

> At a very late stage of cortical evolution, two major developments took place: the emergence of language and the rapid ascent of the executive functions… language acquired its place in the neocortex by attaching itself to various cortical areas in a highly distributed way. And the executive functions emerged as the brain's command post in the front portion of the frontal lobe, the prefrontal cortex. The frontal lobes underwent an explosive expansion at the late stage of evolution.[5]

The prefrontal cortex is the section of the frontal lobes lying at the front of the brain and is the centre of the executive functions in the brain. It is involved in managing complex processes like reason, logic, problem solving, planning and memory. Neuroscientists think that through the integration of these multiple processes the prefrontal cortex is a major component of directing intention, developing and pursuing goals and inhibiting non-serving impulses. It is thought that the prefrontal cortex plays an important role in determining your personality.

Under stress your prefrontal cortex doesn't function optimally as its structure and function can be altered by your experiences. When it is not working well you will feel lazy, lethargic, easily distracted and forgetful. When it is in optimal condition you experience intentional awareness, good attention span, you can plan ahead, focus and think about possibilities for the future.

Inferior prefrontal cortex

The inferior prefrontal cortex is involved in inhibiting previously rewarded responses. It is predominantly involved in cognitive rather than emotional areas and plays a role in weighing up the pros and cons of options without emotional input.

Orbitofrontal cortex

The orbitofrontal cortex is the area of the prefrontal cortex that sits just above your orbits or eye sockets. Located right at the front of your brain it has extensive connections with sensory areas as well as limbic system structures involved in emotions, memory and reward. It is thought to play higher order cognitive roles like decision-making.

Dorsolateral prefrontal cortex

The dorsolateral prefrontal cortex is known for being involved with long-term memory and linking events together through association, particularly memory of information that is personally meaningful. It becomes active when you make a rational, economic decision and is thought to be involved in self-control.

Ventromedial prefrontal cortex

The ventromedial prefrontal cortex is involved in decision-making and is active during guessing tasks. It is also linked to empathic thinking and its activity is reduced when you feel unfairly treated. As it is connected with brain areas involved in emotion such as the amygdala, it is thought to be associated with gut feelings.

The limbic system

As opposed to the highly evolved executive prefrontal cortex, the limbic system is an ancient collection of structures located deep inside the brain often referred to as the seat of your emotions, or the feeling brain. It is responsible for senses such as desire, excitement, fear, anxiety and memory. It is made up of four main areas, the amygdala, the hippocampus, the cingulate (particularly the anterior cingulate) and the insular. As mentioned above, there is some difference in groupings between academic papers and some scientists would include the hypothalamus within the limbic groupings. As parts of the brain rarely function without interaction from multiple regions, however, it isn't a problem from a functional learning perspective where they are grouped.

Amygdala

The almond-shaped amygdala is a collection of nuclei found deep within the medial temporal lobe. You have one in each hemisphere and they are so called as the word amygdala means almond in Latin.

If you are exposed to a frightening stimulus the amygdala is immediately activated, sending signals to areas of the brain like the hypothalamus to trigger a stress response for your survival. It is involved in working out if it is safe to act. It also has a role in the formation and consolidation of memories involving emotional events. It is active not only when you are experiencing or remembering something aversive but also during positive experiences. The olfactory bulb which provides your sense of smell is connected to the amygdala, which is why smells are often strong triggers for memories.

The amygdala is involved with different aspects of perceiving, learning and regulating emotions. It responds to stimuli that may be picked up consciously or unconsciously and is specifically concerned with motivationally relevant stimuli such as fear and reward. Anxiety is mainly mediated by the amygdala and if you are depressed you will often have higher activity.

Hippocampus

The hippocampus is found deep in the brain in the temporal lobe adjacent to the amygdala. There are two hippocampi, one in each cerebral hemisphere, so called from the Latin word for seahorse as they resemble equine fish. This brain structure is thought to have a critical role in memory consolidation turning short-term memories into long ones. Without a hippocampus you couldn't form new memories and it is also central to context-dependent memory. Shrinkage of the hippocampus is particularly severe in Alzheimer's disease. It is also thought to be important in spatial navigation orientation as the hippocampal neurons encode information about your environmental context to create a cognitive map of your surroundings.

Insula

The insula is a small region of the cerebral cortex located deep within the lateral sulcus and separates the frontal and parietal lobes from the temporal lobe. It is activated in many different circumstances enabling your concept of bodily and self-awareness from sensory responses such as feeling hot or cold to feeling empathy for others. It has a role in your perception of pain and increased insular activity makes you hyperaware of any problems in your body. It also plays a part in your response to emotions, your mood and dampening non-serving reactions.

Cingulate cortex

The anterior cingulate cortex wraps around the head of the corpus callosum. It has connections with many other brain regions and a number of diverse functions related to emotion, including assigning emotions to internal and external stimuli, vocalising emotions or desires. It weighs up conflicts between subconscious autonomic and conscious cognitive processes and is part of the automatic control systems for things like blood pressure and heart rate. A primary role of the anterior cingulate cortex is in error detection and correction in addition to rational cognitive functions such as controlling impulses and anticipating rewards. The anterior cingulate cortex has particular types of neurons called spindle cells which have only one dendrite each. They are thought to be important in human intelligence levels as they are linked to speedy information processing.

The posterior cingulate cortex lies just behind the anterior cingulate and is thought to be activated by autobiographical memory recall and emotional stimuli. It is considered part of the 'default mode network' which is a group of brain structures that are more active when you aren't involved in anything requiring external attention, such as daydreaming or recalling memories.

Midbrain

The midbrain plays a role in many of your physical actions together with hearing, visual information processing, eye movement control and mood regulation. Here, I have first discussed the tectum followed by a number of basal ganglia structures – the striatum, caudate and putamen, nucleus accumbens, ventral tegmental area and the substantia nigra. Some of these structures are actually based in the forebrain but are described in their group of basal ganglia within this midbrain section.

Tectum

The tectum is the dorsal portion of the midbrain that is composed of the superior and inferior colliculi. These are rounded bulges that are involved in visual and auditory reflexes. The superior colliculus processes visual signals and relays them to the occipital lobes. The inferior colliculus processes auditory signals and relays them to the auditory cortex in the temporal lobe.

Basal ganglia structures

The basal ganglia are a group of structures found deep within the cerebral hemispheres and are split between the midbrain and the forebrain but described in this midbrain section for ease.

Striatum

The striatum is one of the principal parts of the basal ganglia and is divided into dorsal and ventral sections. The dorsal striatum towards the back of the head contains the caudate and putamen, while the ventral striatum at the front contains the nucleus accumbens.

Caudate and putamen

These parts of the striatum are responsible for the initiation of movement and are two of the main input areas for the basal ganglia. The caudate originates eye movement and the putamen movement in the rest of the body. If you are depressed, you experience fatigue due to decreased dopamine activity in this key area for movement. The caudate and putamen receive the bulk of incoming fibres from the cerebral cortex, with some from the substantia nigra and the thalamus. The fibres from the cerebral cortex often carry information about movement, which is then modified and sent back to the cortex via the thalamus to be put into action.

Nucleus accumbens

The ventral striatum is the part of the striatum closest to your face containing the nucleus accumbens, a part of the brain hugely studied for its role in rewards and impulses. It is associated with the progression from reward drive and reinforcement to being compulsive in seeking reward as part of an addiction. Dopamine is released in the nucleus accumbens whenever you do something rewarding and exciting.

There is a nucleus accumbens in each hemisphere situated between the caudate and putamen.

Dopamine in the nucleus accumbens rises in response to both rewarding and aversive stimuli. If you are depressed then dopamine activity falls and you will likely feel that life just isn't enjoyable any more.

Ventral tegmental area

The functions associated with the ventral tegmental area are diverse, but this dopamine rich area of the midbrain is best known for its part in motivation, reward, and addiction. The involvement of the ventral striatum in reward is most often associated with fibres that travel to the nucleus accumbens from the ventral tegmental area. This pathway is called the mesolimbic dopamine pathway, activated during rewarding experiences and considered a primary component of the reward system. The ventral tegmental area is also thought to be involved with various cognitive and emotional processes but more definitive research is required.

Substantia nigra

Fibres that leave the striatum mostly travel to the basal ganglia components of the globus pallidus and substantia nigra (from there, the fibres extend to the thalamus and other areas; projections from the thalamus carry the information back to the cortex). The substantia nigra gets its name because it is a rich source of the neurotransmitter dopamine which turns black in post mortem tissue. Dopamine is essential for the control of movement and as well as its role in controlling voluntary movement, the substantia nigra is also thought to contribute to a number of other functions and behaviours, including learning, action selection, mood regulation and addiction.

Hindbrain

The midbrain and hindbrain together compose the brainstem. The brainstem is the stalk that extends from your brain to meet the spinal cord. Motor and sensory neurons travel through the brainstem allowing for the relay of signals between your brain and the spinal cord.

The hindbrain is the most primitive part of your brain, is located at the base of the skull just above your neck and is made up of three main parts – the medulla oblongata at the bottom of the brainstem, the pons just above it and the cerebellum at the back. The hindbrain assists in the regulation of autonomic functions, maintaining balance and equilibrium, movement co-ordination, and the relay of sensory information. These are the most basic functions for survival, supported by the oldest parts of the brain.

Medulla oblongata

The medulla oblongata is involved in many of the automatic behaviours that keep you alive such as breathing, blood pressure and heart rate. As well as providing a control system for the automatic nervous system, it also controls involuntary reflexes such as coughing, vomiting and swallowing. Its axons cross from one side of the brain to the other as they descend the spinal cord which explains why each side of the brain controls the opposite side of the body.

Pons

A primary function of the pons is to connect the forebrain with the hindbrain for relaying sensory information. Pons means bridge in Latin and it looks like a rounded bridge just above the medulla connecting it to the midbrain. The pons helps control your autonomic functions, as well as sleep and arousal states. It regulates deep sleep, reducing movement during sleep. Interestingly, the pons is also implicated in your facial expressions and eye movements.

The cerebellum

Cerebellum is Latin for little brain and it looks like a bit like a smaller version of your brain with a distinctive rippled surface. It protrudes from the back and bottom of the cerebral cortex, is densely packed with neurons and best known for its part in modulating movement. The cerebellum relays information between muscles and areas of the cerebral cortex that are involved in motor control. It is richly supplied with sensory information about the position and movements of the body and can encode and memorise information needed to carry out complex fine motor skills and co-ordination. It enables you to move in a smooth manner while keeping you balanced and maintaining equilibrium. Learning movements through practice is partly due to strengthening synapses in the cerebellum.

There are estimated to be 86 billion neurons operating in the brain. The cerebral cortex has around 16 billion of them (19% of the total), the cerebellum 69 billion (80% of the total) and the rest of the brain 1 billion (1% of the total).[6] So the parts of your brain that provide your instincts and the automatic functions that reside outside of your conscious awareness and that served your ancestors well 4 million years ago still account for around 80% of your neurons. The cerebral cortex, particularly the frontal lobes, may have achieved the accolade of 'The Conductor of the Orchestra', but it sure has a large audience, four times its size.

Neurotransmitters

Neurotransmitters are the body's chemical messengers used by the nervous system to send signals between neurons and other cells in the body. They affect a wide variety of both physiological and psychological functions such as your appetite, sleep quality and mood. Neurotransmitter molecules work constantly to keep your brain in good working order, managing everything from your blood pressure to your attention.

When an electrical signal reaches the end of a neuron, it triggers the release of small sacs called vesicles that contain the neurotransmitters. These sacs spill their contents into the synapse, where the neurotransmitters then cross the synaptic gap to the next neuron.

The neurotransmitter then attaches to the receptor site on the receiving neuron, either exciting or inhibiting it depending on what the neurotransmitter is. I like the metaphor that neurotransmitters act like a key and the receptor sites act like a lock. If the neurotransmitter is the right key for the receptor lock, then the changes will occur in the receiving cell.

A neurotransmitter influences a neuron in one of three ways: excitatory, inhibitory or modulatory. An excitatory neurotransmitter promotes the generation of an electrical signal called an action potential in the receiving neuron and is said to have an excitatory effect on the neuron. This means it increases the chance that the neuron will fire an action potential. An inhibitory neurotransmitter has an inhibitory effect on the neuron, decreasing the chance that the neuron will fire an action potential. Whether a neurotransmitter is excitatory or inhibitory depends on the receptor it binds to.

Modulatory neurotransmitters or neuromodulators are capable of affecting a larger number of neurons at the same time. They regulate populations of neurons and operate over a slower time than the excitatory and inhibitory types.

Some neurotransmitters, such as acetylcholine and dopamine, can create both excitatory and inhibitory effects depending upon the type of receptors that are present.

Major neurotransmitters and their roles

Acetylcholine	Mainly excitatory	Primary neurotransmitter associated with motor neurons. Plays role in peripheral, automatic and central nervous systems. Affects arousal, attention, memory, motivation, muscle movement including gastro-intestinal muscles. Too much results in spasms. Too little in muscle weakness, fatigue and even paralysis.
Dopamine	Both excitatory and inhibitory	Essential for memory and motor skills. Regulates reward and pleasure centres – hence role in addiction. Too much can cause hallucinations and delusions. Too little results in poor memory, concentration and attention, no energy. Too little in the motor area can lead to Parkinson's disease.

Adrenaline (also called epinephrine)	Mainly excitatory	Affects arousal, wakefulness, learning, memory and mood. Part of stress response for survival. Too much leads to 'tired but wired', sense of urgency, trouble sleeping.
Noradrenaline (also called norepinephrine)	Mainly excitatory	Primary neurotransmitter in sympathetic nervous system playing a role in blood pressure, liver function, heart rate etc. Affects arousal, wakefulness, learning, memory, alertness, decision-making and mood. Mobilises brain and body in response to stress or danger. Too much can result in anxiety and insomnia.
Gamma-aminobutyric acid (GABA)	Inhibitory	Acts as a brake on excitatory systems. Involved in vision, motor control, anxiety regulation. Too little can cause anxiety, racing thoughts, distraction and in certain parts of the brain, epilepsy.

Endorphins	Inhibitory	Inhibit transmission of pain signals and promote euphoric feelings, appetite modulation and immune response enhancement. Endorphins are made in the brain, spinal cord and other parts of the body.
Serotonin	Inhibitory	Important for sleep onset, memory, learning, regulating mood, sexual activity, appetite. Too little can lead to depression, sleep issues, lack of confidence, feeling panicky.
Glutamate	Excitatory	Most abundant excitatory neurotransmitter in the brain. Plays role in learning and memory. Too much can cause excitotoxicity, killing neurons associated with stroke, epilepsy and brain injuries.

Histamine	Excitatory	Plays a role in allergic responses, metabolism, temperature, hormone regulation, sleep-wake cycle. Too much can trigger allergic reactions, sensitivity to touch, feelings of being hot and a low pain threshold.
Adenosine	Modulatory	Suppresses arousal and promotes sleep.

Scanning technology

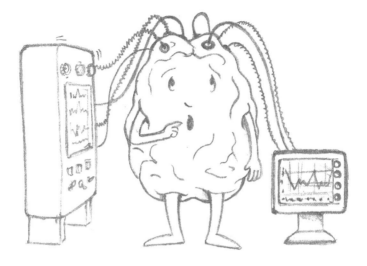

Much of the research in this book wouldn't be possible if it wasn't for advances in technology enabling neuroscientists to take a peek inside the human brain. Scanning methods are improving all the time, as the technological revolution continues to warp speed on, offering scientists tools with ever increasing accuracy and predictive abilities to examine brain activity. Inevitably these scanning tools come with some limitations, and by no means can scientists claim results from either their research or the machines to be 100% accurate and prove causal relationships in all their presenting hypotheses.

As you will see from some of the remarkable studies reported in this book, we certainly live in exciting times, exploring and discovering new insights into the astonishing 1.5-kilogram mass in our heads. In neuroscientific research, readings are made during the task being investigated – and sometimes before and after as well. Hypotheses can then be tested about how brain activity is being affected by certain tasks or behaviours. It is this scanning technology that has allowed us to enter the amazing world of our brains. Here is a short explanation of the various scanning techniques used throughout the studies highlighted in the following chapters.

Functional magnetic resonance imaging (fMRI)

fMRI neuroimaging techniques have been used since the 1990s and are based on the premise that oxygenated and deoxygenated blood have different magnetic properties. When neuronal activity increases there is an increased demand for oxygen increasing the blood flow to that part of brain.

The fMRI scanner works by measuring changes in blood oxygenation and flow that occur in response to activity in the brain. The machine is designed to examine the relative concentrations of oxygenated and deoxygenated blood in small areas of brain tissue and involves the patient lying down, keeping his head still (maybe held in a brace) and sliding into a large cylindrical tube where a strong magnetic field and radio waves are applied to examine blood flow. This produces what is known as a BOLD (Blood Oxygen

Level Dependant) response. If an fMRI scan reveals an increase in the level of oxygenated blood in a particular region, the assumption is that the region is active and requires additional oxygen for fuel. fMRI machines can be used to produce activation images with various colours representing the relative strengths of the BOLD data, showing which parts of the brain are involved in particular mental processes.

Positron emission tomography (PET)

PET scanning uses a dye containing tiny amounts of short-lived radioactive material to show activity in the brain. The scan captures images of brain activity after the radioactive material has been absorbed into the bloodstream usually via an injection. The material attaches to compounds such as glucose, the principal fuel for the brain. Active areas in the brain use glucose more quickly than less active areas and when the radioactive material breaks down, positrons are emitted, creating gamma rays which are detected by the scanner. These rays are analysed by a computer creating an image map of the activity within the brain.

The PET scan is performed lying down, requiring the participant to remain still. The table slides into a large cylindrical tube where the readings are made.

Electroencephalography (EEG)

An EEG measures the electrical activity of the brain via small electrode sensors attached to the scalp and connected by wires to the recording machine. The traces represent an electrical signal from a large number of neurons sending messages to each other, tracking and recording electrical patterns. The EEG is capable of detecting changes in electrical activity in the brain on a millisecond-level.

Functional near infrared spectroscopy (fNIRS)

fNIRS is an optical brain imaging technique for measuring oxygenated blood in the brain. By measuring changes in near-infrared light, it allows researchers to monitor blood flow in the front part of the brain. Differences in the absorption spectra of oxygenated and deoxygenated blood allow changes in haemoglobin concentration to be measured through the light attenuation at different wavelengths. The participant wears a sensor on the forehead which is connected to a computer that records data. This is then analysed for changes in the blood flow and oxygenation levels.

PART 2

COACHING STORIES

PART 2

COACHING STORIES

CHAPTER ONE

Old Habits Die Hard

'A man who can't bear to share his habits is a man who needs to quit them.'
– Stephen King, author, *The Dark Tower*

How much of your day is routine? When I ask this question to a coaching client or a training group the answers couldn't vary more widely – from 1% to 99%. Just think about it for a minute. You wake up in the morning, you hear the alarm going off the same time as it does every day, you press the snooze button and relent after the same number of reminder alarms, you haul yourself out of bed and your feet will probably settle on the same indented part of the carpet. When it comes to brushing your teeth, have you ever looked in the mirror and wondered whether to select a molar or an incisor to start? Most likely not.

A team from Texas A&M and Michigan State Universities in 2002 found that more than 40% of the actions people performed each day in the same location weren't due to making conscious decisions but were habits.[1] Your neurological activity gets less as the habit locks in because your brain is following an established routine.

Using routines and automated processes is a strategy to free up cognitive capacity so you can focus on your goals and engage your intentional, rational mind – what is referred to as conscious thinking. Without the morning habits described above it is unlikely you would get out of the house in time to get to work.

Habits can be broken, changed, or replaced. Awareness of them is a first step which enables you to ask yourself if they are serving you well or not. And simply understanding how habits work makes them easier to identify and manage – a coaching conversation can catalyse this thinking.

When your rational mind is engaged, you act in ways to reach your goals and usually are aware of your intentions. However, your habits function largely outside of your awareness.

Duke University researchers asked participants to taste popcorn, and as expected, they preferred fresh popcorn to the stale version that was offered.[2] But when they were given popcorn in a cinema, people who have a habit of eating popcorn at the movies ate just as much stale popcorn as participants in the fresh popcorn group. Participants ate out of habit, regardless of freshness, when in the context that they associated with previous popcorn eating – the cinema. The thoughtful, intentional mind is easily derailed and 40% of the time they are falling back on habitual behaviour.

PETER'S STORY

Peter was living a life of stale popcorn with little awareness at the time. He came to me recommended by a colleague who I had coached in her previous company. He worked in a leading telecommunications organisation, recently promoted to manage a finance department of 15 people, reporting to the Chief Financial Officer. Before the formal coaching commences, I always have an initial meeting with the prospective client so they can meet me and assess if they want to work with me as their coach. During that session, Peter came across as enthusiastic about his new job, measured and quietly confident about his abilities, busy but motivated to spend time to gain some insights so he could work on his personal development. He was completely aghast about recent feedback he had received from a 360-degree intervention he had participated in for a Leadership Development Programme. The survey was designed around the company's leadership competencies and offered rich data about what the company expected from its leaders by helping participants identify strengths and areas for development. Although anonymous, Peter was able to read verbatim comments his team, and others he had selected to complete the survey on his behalf, had made about him.

He was mainly concerned about scores and comments in two sections of the report – firstly his effectiveness as an empowering leader and secondly his presence as an inspiring leader.

Comments in the area involving empowering others included 'Peter interferes all the time with my work. It's not that he takes credit, he just won't let go', 'I don't think Peter trusts me to do my job, he's helpful and supportive but he needs to see I am capable of running the whole project', 'Peter is a good manager as I am clear what I have to do, but he is always looking over my shoulder', 'I think Peter is a micromanager, he says he will delegate things but he rarely does', 'I don't feel I am developing as Peter doesn't let me suggest new ideas', 'I don't think Peter would know what empowerment was if it flew up and hit him in the face'. Peter told me he was stunned at how harsh some of the comments were; he realised he had a problem letting go of some of the work he was an expert in but it was a shock that there were so many consistent comments in this area. Through our coaching I discovered that Peter had been promoted to the manager position as he had been a top performer in his previous accounts role and it was the expected next step. He was proud of the promotion and six months prior had attended a course entitled 'Managing People' where he had learned about motivation, communication, performance management, giving feedback, coaching, delegation and teamwork. The MyBrain International profiling tool MiND revealed he had preferences for analytical and logical thinking, also enjoying detail, planning and organising (see Appendix for details of the MiND tool). Taken together I describe him to exhibit left brain processing preferences. He admitted he found it hard to delegate, but the issue was not around trusting his team, rather around what the real work of the manager was, saying, 'If I don't get involved with the team to help them do their work, what do I do?'

Discovery conversations around the second red flag from the 360 tool were revealing. Team comments included 'I would like to hear more about the company strategy directly from Peter, his team meetings are all about tasks', 'Peter is a good manager but doesn't show enough confidence to be inspiring', 'I think Peter needs to learn to be confident presenting', 'He needs to engage us as a team more and tell us what is happening in the company', 'I consider Angela to be our leader, not Peter' (Angela was the Chief Financial Officer).

Peter told me he found it challenging to step up to the position of manager for his former peer group and he struggled with the area of being an inspiring, encouraging leader even more than empowerment. He was more task focused by nature and the transition to a leadership role was proving a stretch; he was wondering if he was up to it. The good news was that he approached the coaching with enthusiasm and gusto and assured me he viewed the Leadership Development Programme as an opportunity as well. He had proactively asked for a coach when Angela had provided him with the initial 360 feedback which he told me she had delivered in a professional, non-critical way. However, as someone with left brain preferences, although open to change, he was more comfortable with following a process or a guidebook, rather than dealing with ambiguity, uncertainty and the 'softer' areas of emotions. I knew the scientific evidence offered by neuroscience would appeal to his fact-based preferences so the conversations about what was happening in his brain supported by research were useful. Despite his genuine desire for change, the trust he invested in the coaching from the beginning and the hard work he was prepared to do, Peter was stuck and unsure what his next steps were.

Alongside other discussions, we looked at the neuroscience of habits and how Peter could use this knowledge to help him create more effective habits in the two key areas we identified. We focused on developing a habit of proper delegation, supporting but not micro-managing. Secondly he worked on starting a habit of regular face-to-face team communications about the business bigger picture rather than just about tasks as a start point in addressing the executive presence feedback.

The brainy bit

Even though your habits largely function outside your awareness, they can be changed. This isn't just about willpower or mind over matter but due to a wonderful power in the brain called neuroplasticity. Throughout the whole of our lives the synaptic network is constantly changing as we experience the world around us and learn new information or skills.

Neuroplasticity is lasting change to the brain throughout an individual's life course. It is observed at multiple levels, from microscopic changes in individual neurons when trying a new activity to larger-scale changes such as cortical remapping in response to brain injury.

Neuroscientist David Eagleman in his seminal book *The Brain: The Story of You* talks about neuroplasticity as being like a computer that reconfigures its own circuitry, saying that the brain of an adult, while not as flexible as a child's, still retains an incredible ability to adapt and change.[3] At its simplest, every time we learn something new, the brain changes and neurons can regenerate. No matter how old you are, you still have the power to change your brain and improve your life.

Synapses communicate with each other to form connections – the same things happen with habits. Every time you learn new information or have a new experience you make a new synaptic connection. When you repeatedly engage those circuits, you experience a habit.

Habits have a recognisable neural signature and neuroscientists are increasingly understanding where they are active in your brain. When you are learning a response, you activate your basal ganglia structures, also involving the prefrontal cortex supporting working memory so you can make decisions. As you repeat the behaviour in the same context, the information is reorganised in the brain. As a behaviour becomes automatic it moves from your intentional awareness in the prefrontal cortex into the striatum, an ancient processing centre deep in the brain in the basal ganglia. The striatum receives dopamine from neurons that help form habits by giving you rewarding feedback. This means the action in that situation becomes easier to repeat next time.

Let's remember we are talking about ancient brain structures at play in the above scenarios. Most of our actions aren't a result of conscious thinking but from routines and habits. You take actions because you've laid down the neural pathway by doing it many times before. You may have a routine for biting your tongue when concentrating or a gin and tonic routine at the end of the working day or a twice press of the snooze button in the morning. This is what was happening with Peter. He had previously performed the role that his team members had. That work was second nature to him, he had been a top performer and he felt valued and a contributor when he did it. Even from the new context of his promotion, he was accessing routines to do work that was familiar.

When you do anything you feel is rewarding like eating a chocolate croissant, having afternoon sex, drinking cold prosecco, running through the woods, dopamine neurons in the ventral tegmental area are activated and project to the nucleus accumbens, raising dopamine levels there. Over time, the nucleus accumbens learns what gives pleasure and how to anticipate it. Dopamine is then released in anticipation of the pleasure which blasts you to act undeterred by any consequences and before you know it, the croissants land in your shopping trolley despite your diet. As you near the wine bar initially intending to go home, it is far easier to persuade you to go 'just for one' than inviting you by email in the afternoon. It's as if your nucleus accumbens can taste the sauvignon blanc as you approach.

Do that a number of times and chocolate croissants will miraculously appear on the conveyor belt as you are paying for your weekly grocery shop and the wine bar will become your regular Friday night haunt. The nucleus accumbens motivates the initial behaviour, learns what gives the reward, then increasingly anticipates the reward and may even start craving it. Simply put, if your brain likes the reward, it will crave it, want it, seek it, travel around the neural pathways to persuade you to act to get it. But the brain doesn't differentiate good from bad habits. The habit of checking you have your keys when you leave the house (good habit) is controlled by the same part of the striatum as the habit of checking your phone every time you hear a ping (probably bad habit!).

In his eye-opening book *Hooked: How to Build Habit-Forming Products*, Nir Eyal explores the world of habit-forming product design where companies make their goods indispensable.[4] They attach their product to internal triggers and as a result consumers show up without any external prompting time and time again. Eyal explains that rather than spend hefty marketing dollars, habit-forming companies link their services to the user's daily needs and emotions. Before you can engage your rational brain, if you feel a tad bored, you grab your phone and check your Twitter feed, if you feel a bit lonely, a quick look at Facebook will instantly connect you to your friends, a twinge of hunger gets you ripping open that lovely purple wrapper of Cadbury's milk chocolate. And so on.

Learning from drug abuse

A paper from RTI International, Maryland uses some helpful language when examining substance abuse disorders, saying that habits are a vital component of what we refer to as 'human nature'.[5] They offer a distinction between habit capture (forming the habit) and habit maintenance (sustaining the habit). Habit capture in the brain is dominated by learning processes and controlled by motivational processes of the dopamine influenced parts such as the ventral tegmental area and the nucleus accumbens. They say that experimental drug use is captured by this system and for some people, drug abuse becomes an established and probably irreversible habit.

Habit maintenance operates differently in the brain, largely divorced from any motivational processes and controlled mainly by the basal ganglia striatum systems for execution and actions such as seeking drugs and self-administering drugs. If the habitual behaviour is blocked – if for instance an addict can't get access to drugs – then craving switches the brain back to habit capture mode. In drug use, when there is a transition from voluntary drug taking to more habitual and compulsive use, control moves from the prefrontal cortex deliberative choice to the striatum and there is also a move from ventral to more dorsal parts of the striatum.

The researchers call this transition from habit capture to habit maintenance 'passing of the baton' in the brain of drug abusers.

Once habits are ingrained in the dorsal striatum they don't require the nucleus accumbens to motivate them – it doesn't matter if it is pleasurable or not. The striatum isn't rational or conscious so bad consequences don't matter to it. Addictions usually begin as immediately rewarded enjoyable urges drowning in the dopamine joy in the nucleus accumbens. Eventually the nucleus accumbens stops responding and the addictions no longer feel so enjoyable. But because they are now ingrained in the dorsal striatum, you feel compelled to have another drink or cigarette anyway or to butt in and do your old job like Peter was doing as a newly promoted manager.

Insensitive to the consequences

Very simply, what distinguishes habits from goal-directed behaviour is that they become insensitive to potential consequences. The first time you go to a bar with friends looking forward to your first ever glass of vodka, you may enjoy it (thank you nucleus accumbens) but recognise that feeling tipsy and the resultant fuzzy mind or loose tongue may have a consequence (thank you prefrontal cortex). Keep at the vodka and soon enough you will worry less about the downside of a morning hangover and being late for work. It is the same with behaviours, it becomes so normal for you to behave in a particular way such as being dismissive of junior colleagues, being snappy with your children, allowing yourself to be distracted by social media, that it becomes a routine and just what you do. It often takes a piece of strong feedback or a particularly repugnant event to jolt your awareness into your consciousness for you to realise you have a bad habit and then to consider doing something about it.

Habits are formed because every action activates a specific neural pattern in the dorsal striatum and every time you follow that same pattern the neurons in the dorsal striatum communicate more, becoming wired together more strongly. Each time you travel the same path the pattern becomes more familiar, more easily accessible and easier to activate that neural signature again. The brain wants to follow its familiar, well-trodden paths. Once that path is laid down, it is available for you to access whenever you need it. Swimming or riding a bike after a number of years soon comes back as you enter the neural pathways laid down as a child.

Initial motivation to habit formation involves different striatum areas

Massachusetts Institute of Technology researchers showed in a T-Maze task with rats that the ventral striatum, where the nucleus accumbens is located, is necessary for initial learning of motivated behaviours that may become habits after repeated use.[6] The dorsal striatum then becomes critical.

Different parts of the dorsal striatum have different responsibilities

The dorsomedial striatum and the dorsolateral striatum belong to distinct basal ganglia networks, that scientists now know mediate actions and habits, respectively. A 2015 study from the California Institute of Technology and University of California, Irvine, asked participants to engage in a simple binary choice task and maximise the monetary rewards offered.[7] They were given detailed instructions that each trial would begin by seeing one of two possible initial stimuli (Sanskrit symbols) and they would then take one of two possible actions (selecting a blue circle or red square), which result in the possibility of two outcomes of either a high ($8 to $12) or low ($2 to $6) monetary reward.

The initial stimulus determined the resulting action-outcomes therefore indicating which of the two actions was highly rewarded on a given trial.

Not surprisingly, as the task was simple, participants quickly achieved a high level of performance in each trial condition, selecting the action that yielded the high money distribution in most cases.

Using fMRI scans at different stages of the trials, the researchers examined specific brain areas to find out where information about the action is represented during the initial decision period (stimulus-response) and also where information about the goal of an action is represented. They found that the dorsolateral striatum contained information about responses but not outcomes at the time of an initial stimulus, while prefrontal cortex and anterior dorsal medial regions activated in goal-directed action selection contained information about both responses and outcomes. These findings suggest the differential contributions of these regions to habitual and goal-directed behavioural control may depend in part on basic differences in the type of information that these regions have access to at the time of deciding to do something.

Stress turns the volume up higher and you develop coping habits which can pull you out of potential problems by reducing amygdala activity and the body's stress response (see Chapter 2). But bad coping habits are an oxymoron, simply creating more stress and you can easily find yourself in a seemingly never-ending spiral. On top of that, if you intentionally choose to do things you aren't happy with every day then your brain is getting better and stronger at doing them – sometimes when you are wishing you weren't doing them in the first place. For instance, there may be elements of your job that you don't enjoy but you need to do them and you can perform them to a high standard. They form part of your remit and you always like to do well, however you want to be rid of these activities and are feeling stressed by the hopeless, demotivated sense you get when you think about them. The longer you put off having an open discussion with your boss or finding a palatable way to deal with these job elements, the more stressed you are likely to become. You may start drinking gin as a way to cope with a recent bereavement but you get a shock each week when you put out the recycling. Bad coping habits are bad habits in themselves.

Changing habits

You have the power to create good habits. Your prefrontal cortex is what separates you from other animals. It puts the brakes on, allows the brain to pause and take more considered action. Most other animals live their lives only by instincts, impulses and routines but humans can overcome this through deliberate conscious actions.

Once a habit is triggered, the only way to manage it is by engaging your conscious prefrontal cortex.

Remember that the dorsal striatum has no stake in what you want to achieve with your habits, it doesn't care about your results, it just follows their familiar laid down neural pathways. Understanding these pathways is a key step to changing and to creating better habits. In essence, you can break old, non-serving habits by making the new habit stronger than the old one. If you want to change a bad habit, you don't actually get rid of it, it is still available if you want it, but it gets weaker as you create new, stronger ones. The journey can be challenging, but with some mental effort and a strong goal in mind, it can be enjoyable and liberating.

Using existing pathways in a different context

There are several steps to reprogramming a habit. Remember that an existing neurological pathway is far stronger than one that has only been used a couple of times. If you recognise that the habit you want to change takes place at the same time or in the same location or responds to the same cue all the time, you can use that existing strong neural pathway and replace parts of it. You need to be completely clear what you are replacing those old non-serving parts with. You then create new pleasurable neural pathways, triggering the release of dopamine which in turn rewards the new activity to lay down as stronger habitual tracks over time. But it needs to be enjoyable for you – if you have never experienced what athletes refer to as a 'runner's high' then the suggestion of a 5k run is unlikely to float your boat or drown you in dopamine as an immediate replacement for quitting your mid-morning pastry habit.

Repetition of the new neural circuit will make it stronger and eventually overtake the old one. But it will take time – there is no magic formula and time-frames differ from person to person, habit to habit, context to context. Last year, for various reasons I decided to stop drinking alcohol. I approached this potentially life-changing move as if it was a project, doing a huge amount of reading, research and talking to experts before I made my decision. Doctors know that within ten days at the most, with no imbibing there is no alcohol physically left within the body. Did that mean that within ten days I had replaced my habit of dinner time wine drinking? Not at all! I approached my reprogramming intentionally, always engaging my prefrontal cortex, keeping my goal in my attention at all times and I deliberately rewarded myself with a desert after dinner whenever I felt like it. This was my replacement at the time of day when I felt my wine habit was most ingrained and I could be most tempted. A desert gave me the dopamine reward to lay down a new neural track for a new habit. I figured at the time if I could quit the booze then I could deal with the sugar habit later. It was a successful strategy for me.

Peter had tried to give company updates to his team during his team meetings in the department confer-ence room but clearly the feedback said it hadn't gone well. Once a month Angela ran a department update for the whole finance group in the coffee lounge on the finance floor. We discussed a change in location for Peter's meetings and he started to use the same area as Angela. His team were familiar with using the coffee lounge as an area for communication and were likely to enter that area automatically with an open mind – already having a neural track associated with it. Importantly, the location was also known to Peter as the communications meeting venue. With that anchor in mind, Peter delivered his updates from the same part of the room Angela used, even standing in the same spot and was pleased his first session went well with lots of interaction and questions. With practice, he lost his nerves, built his confidence and became a more proactive communicator. Just by changing the location, by breaking the pattern of his less successful update sessions, he made a change to his delivery as well, eventually laying down solid new neural pathways. He described the rush of success as being like a first date saying yes to a second. He felt really good, justifiably so, and dopamine was helping that high.

Starting a new habit

Sometimes you need to start from scratch with a new habit – it may not be as logical as replacing wine with desert at a time of day that acted as the cue for the old bad habit. Here you need to find the replacement cue then learn how to act it out repeatedly. In this situation strong motivation is essential and the most powerful way of creating a new habit. If you enjoy the new habit you are creating, it will be easier and quicker to create a new routine. To get there, being clear about the ideal world you want to create, certainty around your goal and using congruent language are essential.

Visualisation

Visualisation is the process of creating a mental image or intention of what you want to happen or feel in reality. Many athletes visualise to 'intend' the outcome of a race or a training session. By imagining a scene of success in detail, the sounds, smells, sights, conversations, touches, emotion, the athlete metaphorically steps into their senses – that feeling of winning or gaining a personal best.

The brain doesn't distinguish between real or imagined when you are creating or strengthening pathways. Visualising the new world created by your new habit, describing what is happening, what you are thinking and feeling, mentally rehearsing it, anticipating it – all of these activities will activate the neural circuit and help attract new neurons to it. This makes it easier to choose that new habit until the neural pathway becomes stronger than the old one for the habit you want to be rid of.

In using visualisation with my clients, I sometimes coach them to act 'as if', just as athletes do. This is an effective technique where you learn to act as if you've already achieved your dream or goal. This helps to instil confidence by putting perceived challenges that lie ahead in a better perspective and programming your mind to respond differently to them. Acting 'as if' helps your posture, your breathing, the sense you project and calms your emotions so you begin to see that what you need to do to reach your goals is manageable. Whenever I speak at a conference, I always feel more comfortable if I have stood in the space prior to my session. This is not just to ensure the logistics are in place, what side I enter the stage, steps I may need to navigate, where the lectern and slide clicker are situated etc, but also so I can see myself in the room and act 'as if' I am already delivering a successful speech for a highly engaged audience. I have performed this mental rehearsal for conference speaking so many times, it has now become a habit.

Part of Peter's confusion stemmed from his uncertainty about what his managerial role was all about. He decided to meet with Angela to share his MiND profile results with her and to explain that he needed some concrete guidance around priorities beyond delivery of his formal objectives (see Appendix for details of the MiND tool). Understanding his preference for detail, Angela helped to break down what she was expecting from him, asking him to present a plan of deliverables for the next quarter. With positive feedback from that meeting, Peter felt more confident in delegating more to his team members.

We practised the visualisation and 'as if' technique, which helped him deal with his reticence of letting go of work that he was used to carrying out himself before the promotion. Not surprisingly, the results weren't plain sailing. Some of his team, despite their feedback on the 360-degree exercise, needed more support than others and Peter had to learn to flex the muscles of his new habit as he established his new neural pathway. Interestingly, Peter told me he would have been reluctant to practice 'soft' techniques like visualisation and acting 'as if', if it hadn't had been for the underpinning neuroscientific evidence. He also added more 'as if' practice into his day prior to his team meetings, by replicating my behaviour at conferences, by standing in any room he was speaking in before the session to ground himself and gain confidence.

Affirmations

A number of years ago as part of an icebreaker exercise in a training course, people were asked to share embarrassing moments. After a small group discussion one participant was encouraged to reveal his story to the plenary group. He was training hard for a triathlon and had found a solution to help with his running struggles, finding he ran faster when he listened to Billy Ocean's famous song 'When the going gets tough'. He told the group that he added three words at the end of the chorus repeating 'I am tough' to himself over and over again. Only he wasn't just encouraging himself internally, rather singing out loud and in a macho, determined, strong manner – and was overheard by a group of ladies on a hen night who chased him down the road. Funny story delivered in its original context, he also told us how the affirmation of 'I am tough' spurred him on to keep going, to overcome a challenging run and to encourage a positive attitude to his least favourite triathlon element.

Reciting positive affirmations or writing them down and displaying them so you see them at your trigger times focuses your attention, gives you pause and activates your prefrontal cortex to take you into a rational space and override the striatum.

Researchers from the Universities of Pennsylvania, Michigan and California, Los Angeles conducted a study of sedentary adults who were fitted with an accelerometer to measure physical activity.[8] They had all completed an exercise to elicit a ranking of personal values (such as friends and family, living life in the moment, religious values). In the fMRI segment of the study, participants in the affirmed group reflected on their top values and those in the unaffirmed group thought about their least important value.

Compared to the other group, the affirmed participants showed increased activity in key regions of the medial prefrontal cortex and the posterior cingulate cortex (part of what the researchers call the brain's 'self-processing' regions) and in the ventral striatum and ventromedial prefrontal cortex (part of the brain's valuation system) when reflecting on future-oriented core values compared with everyday activities. In addition, for a further physical activity intervention, this neural

pattern went on to predict positive changes in participants' sedentary behaviour for the affirmed group. The
researchers say that the results highlight neural processes associated with successful self-affirmation, and
further suggest that key pathways may be amplified
when someone is thinking positively about the future.

The language of affirmation and visualisation

Affirmations work, if the language is right and they are
tailored to you, not generic. As part of her dream of a
smaller body, a friend of mine told me she never used
the term 'loss' when it came to her weight. Human
beings have unconscious perceptions of loss interpreting it as giving something up, which equates to scarcity.
Our survival instinct tells us to avoid scarcity at all
costs. Rather, my friend told herself she was letting
her weight go. In so doing, she was in control, she was
dropping the weight from her body, choosing to let it
go. A subtle difference, but an important one.

Change the language also of 'can't' and 'don't'. 'I can't
do [old habit] any more' creates a feedback loop that
reminds you of being limited and deprived resulting
in that feeling of loss as described above. It implies that
you are doing something you don't want to do and
it is less likely to be successful requiring much more
mental effort in the long run. 'I don't do it any more' is
more empowering, telling your brain you have left it
behind. Remember to add in 'I do this [new habit] now

and I love it'. Your affirmation is far more powerful if you remind your brain you are acting in a better way.

As you progress in your habit change, it is important not to pay old habits any attention – even saying 'I have stopped drinking wine in the evenings' gives strength to the old neural circuit. Ensure your language is congruent with the new world you are creating, not the old world you are leaving: 'I enjoy dessert at dinner' or 'I enjoy running 5k mid-morning'.

Keep an open mind

It was essential when discussing habit change with Peter that he understood the need to be relaxed as well as focused. Being stressed about changing habits is not the best state of mind to springboard from. Stress changes the dynamics of the conversation – when you are calm and relaxed, it is easier to keep your prefrontal cortex focused. But the more stressed you are, the more cognitive capacity is used up and the old habits come to the fore in the nucleus accumbens and the dorsal striatum. That is why you may be achieving great results on your diet until you get into an argument with your girlfriend. Or you may be exercising regularly until an urgent project at work is dumped on you. When you are stressed you usually act out your most deeply ingrained routines or give in to impulsive urges, which you often are just not aware of. Your old habits can pop up to bite you at any time. When Peter was stressed, he felt inclined to grab back some of the projects he had delegated to the team, so he needed to be aware of his stress levels and learn to manage them when the pressure was on.

Ask yourself what the secondary gain of continuing with the habit is. In other words, what are you getting out of continuing to do it? Write down your thoughts and feelings for a few days and see if you can identify the crux of your sustaining habit. This requires courage and honesty and is an area that you can work at in depth with a coach. With Peter, because he initially didn't understand what the work of a manager was all about, he continued to do the work his team members should have been doing. Even though he knew he wasn't delegating well, the secondary gain was that he was still delivering good work and contributing to the business. I explained to Peter that even a change in expectations in a new role is often as much about the unlearning as the learning. That is what habit change is all about.

Accept new habits may take time to embed

Impatience flies in the face of habit changing. Repetition and practice are key and inevitably you will make mistakes. Fortunately, when you want to create a good new habit the prefrontal cortex can override the striatum but only when it is focused. Deciding you want to change is an important first step but you have strong old habit pathways embedded in your neural

circuitry. Don't be upset if you aren't initially successful, telling yourself you are a failure won't help and is
another source of stress. Keep reminding yourself of
your new habit goal, read your affirmations, practice
your visualisation routine, being specific in your resolution to change. If you give up after a slip up – guess
what? You are training your striatum to give up after
making mistakes. Stick to your goals, and don't listen
to the devil on your shoulder.

At the beginning of January every year, articles and
blogs groan with advice about resolutions and habit
changing. I have read numerous different timescales
attributed to new habit formation, twenty-one days
appearing to be particularly popular. The reality is that
it depends on so many elements – your commitment,
your goal, your reasons why, your values, your cues,
your environment, your health, your stress levels, the
habit itself – and so on. Without wanting to join the
groan, a study at University College London's Behavioural Science and Health department gives a best
estimate of sixty-six days – taking the achievement of
New Year resolutions to 6 March.[9] They do however
caution that it is probably unwise to assign a definitive
number to the process and emphasise that the sixty-six
days is not a magic number, rather an average time
taken for new habits to become automatic.

Change your environment

Once habits are stored in your striatum, they are triggered by a thought, a feeling or something in your environment – which is something you can control. Your environment may keep triggering them – so identify any specific environmental cue. Or maybe just randomly change something – break the pattern to make a change. Neuro-linguistic programming calls this a pattern interrupt which helps to change thought patterns and behaviours. Get a new screensaver, wear different clothes, drink coffee rather than tea, move your desk, work in a different room, change your office chair from a blue one to a red one, hang different pictures – even subtle changes can have a great effect.

Changing a habit can be easier when you change your environment first. When Peter was promoted he decided to stay at the same desk as he didn't want to alienate himself from his team or appear to be cocky by moving away from his old team mates. A good intention perhaps but remaining in situ only served to make his transition to the manager role more challenging for him at a personal level. In our first session he voiced the idea of moving to a new desk which was pivotal in helping him become more comfortable and confident in his manager role.

Disrupt your habit sequence

It is thought that the chunking of tasks is an important component in how they become habits. The recognition of tasks in a series over time becomes a habit and changing a task in the sequence can disrupt a habit being formed. Similarly, stopping the initiation of the first task in a sequence of a habit can prevent the follow through. The trick is to identify it. New research suggests that once started, the brain wants the whole routine to run. Neuroscientists from the Massachusetts Institute of Technology have now found that certain neurons in the brain are responsible for marking the beginning and end of these chunked units of behaviour in a sequenced habit.[10] These neurons in the striatum fire at the outset of a learned routine, go quiet while it is carried out, then fire again once the routine has ended. The researchers found that excitatory neurons produced what they called the bracketing pattern at the beginning and end while different neurons, interneurons, activated in the middle of the learned sequence.

This task-bracketing appears to be important for starting a routine and then notifying the brain once it is complete. Once these patterns form it becomes difficult to break the habit. The brain considers the pattern valuable and worth keeping. The researchers suggest that the interneurons prevent the excitatory neurons from starting another routine until the current one is finished, implying that once started, the brain wants to complete the activation of the habit. Even more evidence for the need to identify the triggers and stop the initiation of the first task in the routine. Peter's simple strategy of moving desks disrupted his automatic routines when he first came into the office in the morning which helped him develop the new mindset of being a manager rather than a member of the old team.

Identify the triggers

All habits have triggers, a cue which tells your brain to kick into automatic mode and which habit to use. It may take some time for you to identify the trigger but this can be helped whenever you perform the bad habit, by writing down what you were thinking, feeling and doing before you did the habit, then writing down how you felt, what you thought and what you were doing during the habit and then finally scribing what you thought, how you felt and what you were doing after the habit has been completed.

In the excellent book *The Power of Habit: Why We Do What We Do And How To Change, New York Times* investigative reporter Charles Duhigg examines how habits work through discussions with neuroscientists and offers practical advice on how to break habits that don't serve us well.[11] He explains that of course there is no single magic formula, not least because the specifics of diagnosing and changing the patterns in our lives differ from person to person – and habit to habit – and each is driven by different cravings. He offers a framework as a guide for changing habits as an attempt to summarise the tactics for identifying and shaping habits.

In step 1 'identifying the routine' he discusses the loop of cues (triggers), routines and rewards. Figuring out this habit loop is essential to diagnosing and changing behaviour and it is an area Peter and I spent several sessions discussing until he was sure it was right. Step 2 'experiment with rewards' requires us to identify the cravings that drive the routine bad behaviour. He refers to this stage as data collection, advising small adjustments to the routine so it delivers a different reward. For instance if the habit is to break an afternoon cookie purchase, then buy an apple or coffee instead or go and talk to a friend, or take a walk outside – the purpose of which is to figure out the craving – is it to satisfy hunger or to socialise or to stretch your legs? Duhigg advises noting the first three things that come to mind after engaging in a different routine to force a momentary awareness and help recall. Step 3 'isolate the cue' is a challenging stage as we are continually bombarded by so much sensory 'noise' as your habit unfolds. He suggests identifying categories of behaviours ahead of time to see patterns within the five categories of location, time, emotional state, other people and the immediately preceding action.

Over a period of time it should become clear what the trigger is – maybe it is a particular time of day or a certain person from a particular department coming into your office for instance. The final stage 'have a plan' is about shifting the behaviour by planning for the cue and choosing a different behaviour or routine that delivers the reward you are craving. This involves making new choices and setting what psychologists call 'implementation intentions'.

Commit to the habit change

Deciding to commit to the change and writing it down can help to re-engage your prefrontal cortex on a regular basis – and keep your attention on the new habit. See Chapter 5 for a discussion on decision strategies and setting goals. Figure out some commitment strategies to help. Like affirmations, keep these strategies displayed to remind you of your commitment. Examples include making your habit more challenging to achieve – as easy as not buying chocolate croissants, shopping in a supermarket without a bakery, putting your wine rack in the garage or spending your evening away from food in a different room than the kitchen. Limit your choices by setting new rules such as parking in the furthest available space in the work car park so you walk further, getting off the bus one stop early to increase your step count, switching the Wi-Fi off in the evening to avoid constant social media checks or accessing emails at set times only to stop being distracted. Choices not only give you options and opportunities to engage in your old, harmful habits but they also can overwhelm you and create stress.

In Peter's new life as a manager, developing himself as an inspiring communicator, the very act of scheduling six months of 'update' team meetings and booking them in the team diary was enough to take away any choice of indulging in his old habit of being task focused. In *The Paradox of Choice: Why More is Less*, Barry Schwartz explains that setting rules ahead of time frees us from the constant barrage of choices that deplete our willpower and also helps us become far more productive and less stressed.[12]

I am also a big fan of the Pomodoro technique, a time management tool developed by Francesco Cirillo in the 1980s.[13] It involves committing to twenty-five-minute chunks of time where all you do is work on the task in hand. When the timer rings, you take a short break and mark your work with a tick or a cross, which anchors that you have made progress. By setting the rule of twenty-five minutes, this method forces your attention and focuses your rational mind and prefrontal cortex functions onto the task in hand and has helped me work on my habit of distraction.

Look after your prefrontal cortex

The ideal conditions for changing habits are when your brain is in tip top form. The irony of course is that many of us want to change unhealthy habits that our brains and bodies will benefit from so we are unlikely to be starting from the optimal point. In *The Willpower Instinct*, Kelly McGonigal points out that you only have so much willpower and it can get exhausted by overuse.[14] As you go through your day-to-day activities self-control and stress deplete your resources. To be effective at controlling urges, you need to look after your prefrontal cortex and there are certain things you can do to ensure you are giving yourself the best chance for the mental effort ahead. Simply put, what's good for the heart and the body is good for the brain so you won't be surprised by the advice experts offer. Lowering stress, getting enough sleep, exercise, good nutrition and hydration, practising mindfulness are all essential for optimal brain health. Chapter 2 discusses these in detail in the context of managing overwhelm and stress.

A final word on habits

Coaching with a client motivated to change can often itself represent habit change. Many of my clients, like Peter, are not in the habit of engaging in deep discovery and taking time to build self-awareness. Successful coaching conversations will build new neural pathways for self-reflection and engaging resourcefulness. As coaches we actively move our clients from unconscious habits into the conscious arena, giving them time and space to discover what behaviours and actions are serving them well or not. Starting a new habit can seem hard, requiring intervention by the prefrontal cortex – and that takes mental effort, can be tiring and may not be successful all the time at the beginning. But with practice, the burden of action can shift from being an effort to unconsciously easy.

CHAPTER TWO

The Typhoon of Stress

'The secret of getting ahead is getting started.
The secret of getting started is breaking your
overwhelming tasks into small manageable tasks and
then starting on the first one.'
   – Mark Twain, author

A few months ago, I was testing my fifteen-year-old daughter on her geography revision about weather systems. As I read her notes on tropical storms, their behaviour and the damage they cause, a typhoon, cyclone or hurricane seemed like the ideal metaphor for the feeling of stress. Frequently clients tell me their head is about to explode, they have too much to do, they don't know where to start, they don't have time, there is too much change and uncertainty going on around them, even looking at their 'to do' list makes them feel sick, anxious or panicked – and often they experience headaches, palpitations, shaking, bad tummy, nausea and insomnia. Sometimes tearful, sometimes defeated, sometimes confused, sometimes angry, each of these clients are presenting typical descriptions of feeling overwhelmed that is manifesting in stress for them.

The Health and Safety Executive reveals from its Labour Force Survey that 526,000 workers were suffering from work-related stress, depression or anxiety (new or longstanding) in 2016/17.[1] This equated to 12.5 million lost working days in the UK alone. By far the highest component cause was workload with lack of support, violence, threats or bullying and changes at work also cited.

This chapter describes what happens in the stressed brain and discusses strategies for alleviating stress and overwhelm.

JENNY'S STORY

Jenny came to me very clear that she was too busy and needed to learn strategies to prioritise. She believed her performance in all areas was at risk of falling and she had started to drop some of her spinning plates. She worked in a top management consultancy, was viewed as an expert in supply-chain consulting and graded on the talent register as a high potential with the lure of Partner as the next step up. She had a team of twelve, ran the 'Empowering Women' programme in the company, personally mentored three women across the business, was a spokesperson for the company at major events, regularly chairing customer panels and giving press interviews, often at short notice. Both her personal and team targets had recently been increased with the suggestion that if she exceeded them, then achievement of Partner status was a possible reward. One of her major clients had moved to a competitor representing a loss of 35% of her previous year's results.

When I met her the first time to discuss the possibility of working together, she presented as self-assured, sunny natured, full of energy and ideas and a beacon for human rights. We discussed her goals of wanting to become more organised, cut down on distraction and always saying yes to projects. Her office was comfortable but messy, her desk covered with sticky notes and lists but she assured me she was on top of what she had to do. Her desk was adorned with photographs of family, her kids, parents, she and her husband on their wedding day together with a picture of her running across the finishing line of a race. She told me she was training for her third London Marathon and was finding the increase in work was having an effect on her work-life balance but felt proud to have it 'under control'.

In our first coaching session three weeks later she was almost unrecognisable, spilling more detail from her full to bursting work life at such a rate, I had to work hard to pace her to a calmer state. On top of losing her main client, she had recently received resignations from two team members, her boss had moved to a different department and she didn't like his 'arrogant, up himself' replacement. She then tearfully broke open the floodgates of her personal life, which in her words had boiled over last year, with her husband taking a new job travelling for weeks at a time and requiring her to become a 'single mum' (her words) to their two young children. She had experienced the death of her cousin, her best friend from school emigrating to Asia and the declining health of her parents requiring her to travel a 300-mile round trip every other weekend to help, as their only child. She was fraught with guilt at the possibility of investigating care homes but said she felt unable to make any decisions until she had her head straight.

She told me she was shattered yet unable to sleep properly, tossing and turning at night, drinking too much coffee, was getting headaches and didn't feel like exercising any more. At that point Jenny was exhausted, fragile and scared that all her spinning plates would crash on the floor at once. She was overwhelmed and her brain had accessed its stress response big time.

The brainy bit

Everyone experiences stress in some shape or form. You can recognise when you're in stressful situations – you experience 'brain fog', are forgetful, irrational, emotional and do things out of character. Stress can cause your brain to seize up at the worst possible times, in an interview, a presentation or meeting your future mother in law for the first time. But this is a survival mechanism, when faced with a threatening situation instinct overrides any rational thought and reasoning. The brain is vulnerable and can become sick, just like other parts of the body. It is useful for a client to realise that sometimes when they just aren't firing on all cylinders it isn't that they are doing something wrong, rather they are experiencing a physical or emotional response to the situation they are in.

Jenny realised she was stressed but was in danger of creating a spiral of beating herself up about it, that she should have been able to sort it out and shouldn't have got into a stressful position in the first place. I almost visibly saw her shoulders rise from some weights dropping by just discussing the neuroscience behind stress. It gave her 'permission' to change her narrative from it being her own fault she wasn't coping, to her brain was giving her a message that she needed to take some action.

The stress response

When you are in a situation you perceive as dangerous, your body responds to give you more energy, focus, keeping you alert and tuning you into an emergency response mode. This is called the stress response and when it kicks in, energy is directed to your brain and muscles enabling you to respond to the threat. This is often referred to as the fight or flight response – where you fight for survival or flee as fast as you can from the situation.

In evolutionary terms, this is a remarkable system that has helped our species survive. Thank you evolution and the sabre-toothed tiger. At a potential meeting with the tiger, of course you want every muscle in your body to be working to its peak ability, and your brain to be on high alert to deal with the situation.

Tigers aside, many situations elicit the stress response; the sound of a smashed window in the middle of the night, a gang in a dark street, the prospect of losing your job, the death of a loved one or discovering a cheating partner are all examples. Of course, everything in life is subjective and how they are perceived varies according to different people. A situation that is emotionally stressful for one person may be exciting for another, doing a sales pitch, an important meeting, moving house, meeting your new boss. We all process things differently and attach different meanings to events based on our experiences, interpreting results in different ways. If you feel you have delivered a piece of work at a lower standard than usual then you may feel worried that your competence will be criticised, yet your colleague may just brush it off as a one-off.

There are three main parts of your body responsible for the stress response – your hypothalamus and pituitary, both located in your brain and your adrenal glands, situated one on top of each of your kidneys.

Once your brain decides there's danger, the sympathetic nervous system prepares your body for action and sends immediate nerve signals down your spinal cord to your adrenal glands telling them to release the hormone adrenaline. This increases the amount of sugar in your blood, accelerates your heart rate, raises your blood pressure, and dilates your pupils, bronchial passages and coronary arteries. When in danger you breathe quicker; your faster heart rate and raised blood pressure ensures you pump maximum oxygen and energy-rich blood to your muscles. Your liver releases more sugar into your blood, to release more energy, ready for action. While blood flow to the muscles and organs is prioritised, blood flow to the surface of the body decreases, which explains the phrase 'she turned white as a sheet'. Gastro-intestinal activity also falls which can lead to a feeling of butterflies in the stomach.

The hypothalamus sends signals to the pituitary gland at the bottom of the brain releasing the hormone cortisol. This increases blood pressure in the arteries mobilising glucose from fat tissues. When you perceive that the threat has passed then your parasympathetic nervous system helps to restore your body to a state of equilibrium.

The curse of cortisol

Without cortisol you would die, as it is critical in keeping your blood sugar and blood pressure high to help you escape from danger and optimally cope with life-threatening situations. Your body's stress response is perfectly in tune with cortisol in the short-term but damaging in the long haul. However, some people perceive every day of their life as stressful. The prolonged effect of the stress response lowers your immune system and raises blood pressure which can lead to hypertension and headaches. This can result in weight gain, particularly abdominal fat, and has a role in osteoporosis, digestive problems, hormone imbalances, cancer, heart diseases and diabetes. Adrenal fatigue can result in physical symptoms such as 'tired but wired', weakness, low blood pressure, dry skin, digestive problems, low blood sugar, cravings for sweet, starchy food, feeling faint. And that is just for starters.

If stress is more sustained rather than the short single visit of the tiger then the raised cortisol has many negative effects for your brain as well as your body. It has a domino effect on your mental health and is a disaster for your brain function.

Your brain cells can die

Cortisol released during stress travels into the brain and binds to the many neuron receptors. Through a series of reactions, this causes neurons to take in more calcium, through their membrane. If they become over-loaded with calcium they fire too frequently and die as they become over excited and unable to survive.

Cortisol and BDNF protein

Neurotrophins are molecules that promote the development, health and survival of brain cells. Among the neurotrophins, brain-derived neurotrophic factor (BDNF) is a major regulator of synaptic plasticity and neuron survival – processes that have key roles in memory formation and learning. BDNF is a protein that is central to keeping brain cells healthy and to stimulating new ones to form through neurogenesis. Just like eating protein for muscle growth, BDNF is the protein for your neurons – it's like the ultimate brain super-food. But the raised cortisol of stress stops the production of BDNF so fewer brain cells are formed. Forget super-food and enter a deprived diet.

In the Weizmann Institute of Science, Israel, a team of researchers looked at young and adult rats to see if healthy BDNF levels could help dampen stress and manage the level of cortisol secreted.[2] In the study rats were subjected to four weeks of chronic mild stress to induce depressive-like behaviours. One group had the BDNF reduced and the other increased. The results showed that a reduction of hippocampal BDNF in young, but not adult, rats induced prolonged increases in corticosterone secretion (cortisol is the predominant glucocorticoid in humans, whereas corticosterone is the dominant glucocorticoid in rodents). They also found that hippocampal BDNF plays a critical role in the development of neural circuits that control adequate resilience to chronic stress. In summary, BDNF matters – and particularly so for young rats.

BDNF and aging

It will come as no surprise that like many other chemicals in the human body, aging decreases BDNF levels. That's why it takes you longer to learn to do complex tasks as you get older. Remember, the protein is instrumental in learning quickly and learning well.

Lowered levels of BDNF are also associated with neurodegenerative diseases such as Alzheimer's, Parkinson's and Huntingdon disease although the research is in its infancy to truly understand the cause and potentially create BDNF therapies.

Cortisol and dementia

A Brazilian Study in 2013 of 309 elderly people (from a common genetic background living in the same area) showed that chronic stress and elevated cortisol contributes to dementia in the elderly and speeds up its progression.[3] They found that concentrations of morning serum cortisol are higher in people with dementia, intermediate in people with CIND (cognitive impairment but no dementia) and lower in the healthy elderly control group. The researchers found that the elevated cortisol levels could be a result of hippocampal damage – part of the brain implicated in the regulation of cortisol. Or the results could be that the people with higher levels may have had higher cortisol levels during their lives which could have caused hippocampal decline, possibly triggering the dementia or CIND. As with much of the research in neuroscience, this study offers useful information about cortisol and dementia, but further research is needed to look at causal relationships and how this can affect strategies at different stages of life.

Cortisol and depression

Chronic stress has also been linked to depression. A common feature of depression is the excessive release of cortisol in the blood which can reduce the levels of critical neurotransmitters notably serotonin and dopamine, damaging their receptor sites. Low levels of either of these neurotransmitters can leave you depressed and more prone to addictions. Serotonin is often referred to as the happy hormone, playing a major role in mood, learning, appetite control, and sleep. Dopamine is known as the motivation hormone and is related to pleasure or reward and the drive to seek that pleasure. Too little dopamine can leave you unenthusiastic, demotivated, lethargic and depressed. People low in dopamine often try to give themselves a temporary boost in dopamine levels. You may reach for coffee or aim for a sugar hit to get you over the afternoon slump or alcohol to get you through the evening.

Serotonin-based depression is accompanied by anxiety and irritability, while dopamine-based depression manifests as lethargy and lack of enjoyment of life.

Emeritus Professor of Neuroscience and Cambridge University Jo Herbert, in 2012, in reviewing the relevant studies and literature stated that there is little doubt that cortisol plays a central role in the onset and course of major depression disorder, concluding that the time is now right for serious consideration of the role of cortisol in a clinical context.[4] In the next few years, as well as increased research into the physical rollercoaster that elevated cortisol takes people on, I am sure we will read more exciting discoveries about the implications for our brains.

Stress leads to premature aging

Stress appears to influence the rate of telomere shortening. Telomeres are protective caps at the end of your chromosomes similar to the plastic tip on a tube of superglue. Every time a cell divides, the telomeres get a little shorter. When they reach a critically shortened length they tell the cell to stop dividing. Telomere length has now been linked to chronic stress exposure and depression. A team of researchers across United States universities studied a group of healthy women and found that those with the highest levels of perceived stress and ruminative thoughts had telomeres shorter on average by the equivalent of at least one decade of additional aging compared to low stress women.[5] So at the cellular level, stress may promote earlier onset of age-related diseases.

The prefrontal cortex can shrink

Remember that the prefrontal cortex is the most evolved part of the brain in charge of your highest cognitive abilities. It's also the most sensitive to stress. Even mild stress can cause a noticeable deterioration in your cognitive abilities and prolonged stress exposure will actually cause structural changes in your prefrontal dendrites. Dendrites are the small branches coming off the neuron essential for electrochemical stimulation.

Stress induced dendritic atrophy in the medial prefrontal cortex causes a decline in working memory and increased anxiety is a result of dendritic remodelling in amygdala neurons. Researchers in Korea discuss that structural changes in these areas may be associated with distinct behavioural symptoms seen in stress-related mental disorders.[6] The good news is that this stress induced remodelling in the amygdala and medial prefrontal cortex is reversible, as are the associated cognitive and behavioural outcomes if you can reduce your stress levels.

Emotions and memories

When you are stressed, signals in the brain associated with factual memories weaken while those associated with emotions strengthen. An Indian study found that the growing dominance of activity in the amygdala over the hippocampus during and even after chronic stress may be a factor in increased emotional symptoms, as well as decreased cognitive function, observed in stress induced psychiatric disorders.[7] Stress builds up in the amygdala, the brain's fear centre, increasing the size, activity level and number of neural connections in this part of your brain. As mentioned above, chronic stress causes dendritic atrophy but it also causes loss of spines and suppresses growth in the hippocampus. These are stress effects at the molecular and cellular levels and are thought to impair hippocampal learning and memory. But chronic and repeated stress triggers opposite effects in the amygdala by strengthening the structural basis of synaptic connectivity through dendritic growth and spinogenesis resulting in fear and anxiety-like behaviour. The researchers reported that structural brain changes caused by stress differ not only with spatial location but also over time. Stress is a vicious circle making you more anxious, even fearful.

The bottom line is that stress impedes learning, achieving your goals and relating to others. In addition, the accumulation of stressful events can make it more challenging to deal with future stress. Stress can trap you with cortisol constantly playing dominoes in your brain and body all the time – knock one down and then another and another – you get the picture. And the part of the brain that makes you uniquely human, the superior species on earth – your prefrontal cortex – loses out in the battle for energy when high-stress is involved. In chronic stress, your brain and body aren't getting the time or the resources to regenerate. At work if stress leads to sickness and you take time off, this may lead to more stress, and bingo, more dominoes. In short, stress makes your brain vulnerable. This is what was happening to Jenny, she was in the middle of a dominoes game, with a seemingly endless list of stressors in her life, wobbling at every turn, thinking one wrong step would set the dominoes into a cascade of failure.

Reducing stress

Some degree of acute stress is useful as it readies your brain for an optimal response. Adrenaline release can motivate you to work hard and do well under pressure and in healthy people, once a threat has passed, levels of stress hormones return to the normal homeostatic balance with no long-lasting effects. Even after chronic, long-term stress you can recover. Neural stem cells can regain their ability to regenerate at a normal level when the tiger has moved to another area.

What can you do to stop the effects of chronic stress and cortisol? Managing stress in a useful way comes from making good day-to-day choices and this requires awareness and effort. Your brain wants stability and to minimise uncertainty so you need to have systems, plans and processes in place.

Recognising it is the first step. Obvious symptoms include those already mentioned such as short temper, difficulty in making decisions, tense muscles, difficulty sleeping, headaches, feeling nervous, difficulty focusing, just for starters. Less obvious ones that you may not associate with stress are weight gain around your abdomen, cravings for sweet foods, a need for coffee or another 'pick-me-up', feeling 'tired but wired'. If it has been going on for some time and you have built some resistance to it, then identifying the cause, the trigger, and putting in place strategies may be challenging. Of course, that is where a coach can help. The best way to calm the stress response is to consciously pause and take time to understand the underlying anxiety.

Healthy habits can help you avoid or reverse the effects of chronic stress. To calm down the stress remember that what is good for the body is good for the brain. You can make good conscious choices for your body, thinking about what you eat and drink, how much exercise you get, how many hours sleep you have – and even the thoughts you pay attention to. As well as working hard to calm the causes of her stress, Jenny sought help from a nutritionist friend. Not surprisingly, tests revealed she was suffering from adrenal fatigue as described earlier, which explained her shattered but wired demeanour. A healthy, adrenal boosting diet was prescribed, along with relaxation exercises and yoga. As Jenny progressed in her stress busting journey, she found the yoga, as well as calming her mind, made her running stronger and more efficient.

There are many strategies you can adopt and techniques you can learn to help your brain and to halt the cycle of overwhelm, to stop obsessing about the seemingly infinite amount you have to do. The remainder of this chapter provides details to help you learn to deliberately single task, prioritise what matters to you, make exercise part of your routine and see only the upsides of meditation.

The menace of multitasking

Whatever method you use to plan your time, be happy with the fact that the ability to multitask is a myth. Multitasking suggests doing more than one task at the same time, but what's actually happening is that you're rapidly switching between tasks. That switching makes you feel tired much more quickly than focusing your attention on just one thing. Jenny likened the exhaustion of task switching as like having to reload different programmes on her computer each time she made a change. Working on multiple tasks at once is less efficient because it takes extra time to shift mental gears every time you switch. For most people this is the point when you reach for a pick-me-up biscuit, a cola or a sugary coffee, your focus diminishes, your pattern is interrupted and you may allow distraction to visit.

The prefrontal cortex begins working any time you need to pay attention, helping you to focus on a single goal and carry out the task by co-ordinating messages and information flow to other brain systems. Working on a single task means both sides of the prefrontal cortex are working together in harmony. Adding another task forces the left and right sides of the brain to split the labour and work independently. Scientists at the French biomedical research agency INSERM in Paris asked study participants to complete two letter matching tasks at the same time with the added complexity of matching upper and lower case.[8] The results showed that the brain effectively splits in half – activity in the left side of the prefrontal cortex corresponded to one task while the right side took over the other task, pursuing its own goal and reward. When adding a third task, participants consistently forgot one of the task details and made three times more mistakes than when given two simultaneous goals.

According to the researchers, for everyday familiar behaviour, dual tasking works – you can cook and talk on the phone at the same time for instance. A problem arises when you pursue three goals at the same time. Your prefrontal cortex will usually discard one. It will however depend on the task and the part of the brain involved in its action. Multi or dual tasking while doing natural tasks like eating and walking or simple tasks like washing the dishes are much easier than more

complicated tasks like texting while driving. Those natural tasks place less demand on the prefrontal cortex,
creating an easier switch between eating and walking
to your next meeting.

Eating for instance involves practised motor skills
which don't overlap too much with other tasks which
involve different parts of the brain such as language or
auditory cues. You could be chatting, eating a meal and
looking at photos at the same time. In reality though,
it doesn't take much to focus more on one activity and
take attention away from another so all of a sudden
you miss a question about the photo or what recipe
you used.

A study at University College London found that
participants who multitasked during cognitive tasks
experienced an IQ score decline similar to those who
had stayed up all night.[9] Some of the multitasking men
had their IQ drop 15 points, leaving them with the
average IQ of an 8-year-old child. And you are doing
this to yourself all the time when you glance at your
devices. Be honest, have you ever hopped onto social
media when doing something else? Trying to listen to
a presentation when checking emails and your Twitter
feed in reality means that little information will be
stored from either task. From the brain's perspective,
social media increases the stimulation and burden.

In the Sackler Center for Consciousness at the University of Sussex, researchers compared the amount of
time people spend on multiple devices (such as texting

while watching TV) to MRI scans of their brains.[10] They found that, independent of individual personality traits, people who used a higher number of media devices concurrently also had smaller grey matter density in the part of the brain known as the anterior cingulate cortex, a region which plays a part in cognitive and emotional control functions.

While more research is needed to determine if this rapid switching is physically damaging the brain, it's clear that doing it has negative effects. The way you are interacting with technology may be changing the way you think and these changes might be occurring at the level of cellular brain structure.

Calm your stress with single tasking

It is challenging to focus on one task at a time. Not multitasking at work can be difficult to stop, especially when there is a lot on your plate. There are a few simple and conscious changes you can make to work more efficiently which may require initial mental effort but which will lower stress and create better results in the long term. Researchers from the Federal Aviation Administration and the University of Michigan have proposed new models of cognitive control.[11] The first, goal shifting, involves actively deciding to change tasks. Once you have decided to switch processes, your brain lays down new habits. As discussed in Chapter 1, this requires your brain to turn off the cognitive rules of the old task and turn on new rules for the next.

A group of researchers from Aalto and Turku Universities in Finland and the Max Planck Institute in Germany measured different brain areas of research subjects while they watched short segments of the Star Wars, Indiana Jones and James Bond movies.[12] Cutting the films into segments of about fifty seconds fragmented their continuity. In the study, the subjects' brain areas functioned more smoothly when they watched the films in longer segments of six and a half minutes. The posterior temporal and dorsomedial prefrontal cortices, the cerebellum and dorsal precuneus are the most important areas of the brain in terms of combining individual events into coherent event sequences. These areas of the brain make it possible to turn fragments into complete entities. According to the study, these brain regions work more efficiently when dealing with one task at a time.

Big rocks chunking

I have always liked the late Stephen Covey's story about 'big rocks', detailed in the book *First Things First* where he discusses the need to differentiate important from urgent tasks in your life as a means of managing your time.[13] It is a great way to challenge yourself by writing down on a two-by-two matrix where you actually spend your daily twenty-four hours. If you are honest with yourself, you could find you are spending hours just messing around on Facebook which during a working day most likely falls into the unimportant and not urgent box. Covey provides a powerful metaphor of this time management philosophy, which you can view him demonstrating on YouTube at one of his seminars.[14]

Covey uses a bucket to symbolise your life, a few big rocks, depicting your important priorities and some small pebbles to represent the urgent, busy, non-important tasks. He starts by pouring the pebbles into the bucket then asks an audience member to fit in the big rocks, which quickly can be seen as impossible. He shows the alternative way; putting the big rocks into the bucket first then asking the participant to pour in the pebbles which fill the cracks between the rocks. Sometimes sand is then used to fill in the smaller gaps and then water representing even smaller and less important activities. The point is that unless you put the big rocks in first, then you won't get them in the bucket and translating to the real world, they either won't happen, or become a source of stress.

I use the big rocks story with my clients so that they understand the philosophy of differentiating between important and not important tasks and where they let time pull them. Simply defining what matters is a useful exercise so you can then challenge the elements you list as unimportant. Do you even need to do them or can you decide to let them go? It doesn't matter if some of the pebbles don't get attended to as long as the big rocks do. Some pebbles, sand grains and drops of water can spill over the edge and your life won't be over. And in the reality of life, of course urgent matters will arise and you can choose how to deal with those throbbing pebbles of urgency. Always remember, when you say yes to something, you are saying no to something else – or turning it into a pebble or a grain of sand.

Train your brain to single task

Rather than bouncing back and forth between tasks every other minute or so, dedicate chunks of time to a certain task. The trick is to learn to prioritise so your big rocks don't fall out of your limited bucket of time.

Write down your big rocks

One of the first things I do with my clients after we have agreed to work with each other is to ask them what colour of journal they want to use for their coaching commitments. I am a firm believer in writing things down, for self-reflection, lists, notes, mind maps or just doodles. Like visualisation, writing acts as a mental rehearsal and starts you thinking about achieving the task you are writing about. Even before the pen hits the paper you are putting some thought into what you are writing. This process also helps recall at a later stage and starts you evaluating the information.

One trap people fall into is to consistently avoid tackling the larger, more major projects as they seem so onerous or they don't know where to start. The best way to overcome this is to break them down into much smaller, achievable chunks. For me, 'write my book

STUCK by the deadline of December 1st' was a pretty big goal whereas 'gather studies on habit change', 'research up-to-date investigations on optimism' or 'talk to three clients about including their story' as sub goals took me forward. Small chunks are far more specific and manageable – actually standing a chance of getting done.

Calm the brain down with a plan

Research shows that tasks you haven't done distract you by preying on your mind, however just planning to get them done can free you from anxiety. The study showed that participants underperform on a task when they are unable to finish a warm-up activity that would usually precede it. However, when concrete plans to finish the warm-up activity were made and noted down, performance on the next task substantially improved. Simply writing the tasks down will make you more effective.

Planning from the good old to-do list

Firstly, write down your entire to-do list in any order it flows out of you. I encourage my clients to take no more than twenty minutes, otherwise they may add items that aren't even pebbles or grains of sand.

Next, categorise the list into broad clusters that work for you. In her new sunshine yellow journal, Jenny started with categories of work – client projects, work – woman's group, work – team, then family, friends and exercise.

Now the decisions come. To gain clarity and order around the mass of tasks you need to be honest about your big rocks and pebbles. As well as a journal, I also give my clients a variety of coloured sticky notes, highlighters and a coloured pen set. This part of the exercise asks you to revisit your categorised to-do list and highlight your big rocks – choosing different colours for important/not important and urgent/not urgent items. You will find that more useful categories emerge. Jenny saw clearly that she needed to take a more holistic view of her activities and give more focus to her personal life if she was to take control of the stress rather than the other way around. For her, the category of exercise with her marathon training schedule had become stressful so she treated items on that list as not important, so guess what? She rarely went to her running club, or read the book for book club, or even made the book club meeting.

This is an important stage in the coaching conversation as it yields a far wider and richer conversation than just achieving prioritisation – even though with Jenny that was a major win and I could see her calm down before my eyes. Some people find it hard to distinguish between big rocks and pebbles, believing everything is important. To help with identifying important activities, it helps to get clear on your personal why (see Chapter 6). For some people, identifying goals, vision and purpose up front is a better route to achieving a priority list than the other way around. Indeed, some people also respond well to a structured journal or to-do list with categories pre-defined rather than a plain page.

However, with most clients who present with stress and overwhelm, such as Jenny, I find it is better to encourage them to dump everything going on in their mind on a page (or pages) and structure later. Writing things down means you know the thoughts are out of your head so you stop going around and around in circles which can be mentally exhausting.

Prioritising using the important/not important and urgent/not urgent helped Jenny to work out how she could say no to all the requests she received – both at work and at home. If requests were pebbles or grains of sand then it gave her a clue. If they were important in terms of her fairness and human rights motivation, then she always gave them consideration, even if they weren't in support of her own wider goals.

Once you have created priorities, you can then schedule them by putting them in your diary. Ask yourself which category needs more focus each week or each day, what are the big rock/not important combinations that often fall off the cliff. Activities like going to the gym, taking a break, picking up your kids from school, taking a self-development class, reading a novel. Like me, Jenny enjoys watching TED talks on a variety of unrelated subjects. She told me she watched more on holiday than during her working week. TED talks are deliberately designed to be no more than eighteen minutes and many are shorter. Our conversation revealed that Jenny gained many insights and ideas from this source, always felt sharper, inspired and energised after watching, and frequently used what she learned in her team and client meetings. She reframed her TED talk time as an enormous boulder in her life and put two slots per week in her work diary.

Once priorities are scheduled, you can try different patterns to gain focus and single task. Breaking down the priority into even smaller chunks helps in allocating time. Jenny liked the Pomodoro technique with short bursts and a break (see Chapter 1). Her diary may have blocked out three hours, and she structured that three hours into four or five segments of activity – including breaks, a change in pattern and a discipline about when to access emails or take calls.

In our initial meeting, Jenny told me she was anxious about the possibility of forgetting everything she had to do or losing one of her magic sticky notes. Chunking tasks down, using her journal and creating a categorised list visibly made her feel more relaxed and in control. She was able to make sense out of her jumbled thoughts, stop obsessing and put things in perspective. Most importantly she was able to immediately cross items off her list, learn strategies for saying no and stop trying to own absolutely everything. The simple to-do list was instrumental in that. It dampened her anxiety, helped her to prioritise and acted as a reference point to show her just how much she was achieving. It also had the added effect of enabling her to see that her work life was under control so she could spend real focused time on her home life, her children, parents and her own health through relaxation and exercise.

Strategies to help stay focused on the task at hand

For many people morning can be the most productive time of day. If that is the case for you, tackle the hardest task first thing rather than dropping and revisiting it later. In Dan Pink's recent, highly engaging book *When: The Scientific Secrets of Perfect Timing,* he amasses a huge body of research to show that your cognitive abilities do not remain static over the course of the day and different people have different peak times when it is best to carry out important tasks that require focus.[16] There are larks who do particularly well in the morning and owls who perform better later in the day. Science shows that scheduling and careful timing of your daily routine is crucial to your well-being. Walk away from your desk to disrupt your thinking and gain energy. Breathing deeper, standing up, drinking water, talking to a colleague all help. Pink states that breaks are 'not niceties, but necessities'.

THE TYPHOON OF STRESS

Get away from your distractions

Jenny was an avid Facebook fan, but she recognised that she had almost fallen into the trap of FOMO – fear of missing out. Many of her friends didn't work and she needed to remind herself that she genuinely loved her job, was proud of her achievements and qualifications, and most importantly she wanted to help people every day. She intellectually understood that seeing what her friends were eating at the café wasn't relevant when she was helping clients at work. But she acted on impulse and was forever looking at her phone at work, jesting that she was behaving like a teenager. The simple act of removing her phone and putting it in her desk drawer helped. Making it a bit harder to access turned the prefrontal cortex on as her impulse was interrupted by a conscious thought. As time progressed she developed additional strategies and bought a different phone for work, without downloading the Facebook app. She then went even further and to enable her to sleep better, she turned off her home Wi-Fi at 10pm and invested in a good old-fashioned alarm clock. Positive knock-on effects from laying down new neural pathways and creating new habits (see Chapter 1).

Exercise helps BDNF

Exercising is the fastest way to increase BDNF levels which is why it is so effective in helping to alleviate stress. A University of California study showed that after just two weeks, those who exercised daily produced the protein much more rapidly than those who exercised on alternating days.[17] Don't worry about having to make it a daily habit as after a month, there was no difference in production of BDNF between those who exercised daily and those who exercised every other day. They also noted that the protein returned to baseline (non-exercise) levels after just two weeks of not exercising. The good news is that it shot right back up after just two days of exercising. The message gives strong evidence to get away from the couch and flex some muscles.

And there is more good news. Neurochemicals abound as a result of exercise. Activity increases the firing of serotonin neurons, which causes them to release more serotonin. It also increases the neurotransmitter norepinephrine and the dopamine system. The brain also releases endorphins, neurotransmitters that act on your neurons like opiates by sending a signal to reduce pain or provide anxiety relief. One life change – exercise – can have multiple, seemingly unrelated benefits. Exercise causes loads of unnoticed brain changes by modifying circuits, releasing positive neurochemicals and reducing stress hormones. And here's the thing – you can do slow walking or chair Pilates. No one is asking you to sign up for an iron man. Focus on what you can do rather than what you can't.

Meditation

Mindfulness meditation trains you to put to one side the constant high-speed train of thoughts travelling through your mind. It teaches you to be still, to be calm, to sit in the present without any fear about the future or worry of the past getting in the way of the peace of the moment. Essentially it does the opposite of stimulating your cortisol and stress. It elicits peace and tranquility, bringing balance to the body and brain.

Focusing on the present helps. Worrying and anxiety are projections of yourself into the future, they don't exist when you are fully present. It's all about paying attention to what is happening now, noticing and focusing on the present. Buddhist monks practice mindfulness being aware of the now, just noticing, without attaching any emotional reaction to it. This cuts off anxiety at the source. Taking a pause, a deep breath, inhaling and exhaling slowly, focusing on the breath itself, calms the sympathetic nervous system and reduces stress.

Meditation techniques move your cognitive judgement from threat to challenge, decrease ruminations, and reduce stress arousal. University of California, San Francisco researchers showed that some forms of meditation may have positive effects on telomere length by reducing cognitive stress and stress arousal.[18] Mindfulness increases positive states of mind and reduces cortisol production as a result of stress which may promote telomere maintenance. There is more research needed but it is encouraging initial evidence that meditation positively affects your telomeres.

Carnegie Mellon University researchers, working in a new discipline called health neuroscience, also showed that mindfulness meditation training, compared to relaxation training, reduces inflammation in the body.[19] Brain scans showed that mindfulness meditation training increased the functional connectivity in the dorsolateral prefrontal cortex. Participants who received the relaxation training did not show these brain changes. The mindfulness group had lower Interleukin-6, an inflammatory health biomarker, and the changes in brain function accounted for the lower levels. The researchers believe the brain changes provide a neurobiological marker for improved executive control and stress resilience through meditation. This increases your brain's ability to help you manage stress, and these changes improve a broad range of stress-related health outcomes, such as your inflammatory health.

There are many other stress relieving activities such as getting enough sleep, labelling emotions, connecting with others, and giving back which are just as useful as the ones discussed so far in this chapter. Each of these will be discussed in detail in later chapters relating to other presenting issues clients bring to the coaching conversation. The following are brief outlines of these activities.

Philanthropy

Giving back can give you a huge boost and feel highly rewarding. Just knowing you have contributed in a meaningful way can result in what psychologists call the 'helper's high'. Giving back produces endorphins in the brain that provide a mild version of a morphine high. This also happens when you can truly connect with your sense of purpose, so you know for instance where your work contributes to the wider picture. For a more detailed discussion around finding purpose see Chapter 6 on motivation.

Socialising

Studies show that people with poor social support are less resistant to stress and more likely to have a mental illness. Connecting with other people helps release oxytocin which can counter some of the negative effects of stress. Although sometimes you may feel you don't want to socialise, almost always you will feel better when out and about with people. But it's important that you trust the people you are with and they aren't causing you the stress in the first place. Just having a coffee and telling someone who knows you well what is making you so stressed helps reduce pressure and work out a solution. The old adage 'a problem shared is a problem halved' certainly holds true. For more information around social connection and the effect on your brain, see Chapter 4 which discusses loneliness in detail.

Labelling

A coach can help you to deconstruct what determines you finding things stressful at a particular point in time. Pivotal to that may be how you are labelling things, what you are chattering about to yourself every day. You can then work to change these labels and re-programme yourself through creating a new set of labels. Seeing your state and being able to give it a label can be very empowering. It helps you get a little distance from the stress and that in itself can help reduce it. Labelling is discussed in more depth in Chapter 3 on dealing with negative thinking.

Sleep

Stress and sleep have a two-way relationship. Stress can make sleep more difficult, yet a good night's sleep can help reduce the effects of stress. We all know sleep is essential and I am pleased that the subject is becoming more generally discussed beyond the science lab or the GP's surgery. Sleep improves the health of your prefrontal cortex and is essential for communication across the brain. The hippocampus, essential for forming new memories, functions properly only when you have had a full night's sleep. The more you sleep and the better quality of your sleep, the better your mood, fitness, health and stress levels will be. Yet this is the catch-22. The stress response causes hyperarousal and sleep can be difficult to come by. You may wake up in the middle of sleep cycles, with the stress response triggered by rumination that has resurfaced, then unable to get back to sleep quickly. Good sleep hygiene such as a supporting evening routine can help, as of course can the other approaches advocated earlier in this chapter. For a more detailed discussion around up-to-date findings on sleep, see Chapter 5 on decision-making.

Jenny deferred her London Marathon entry until the April 2018 race which was the hottest on record. She called me to say she completed it in less than four hours and felt that she was a new woman. Her life was organised and fun, she was 'busy but happy busy'. In the early days of our coaching relationship she used language like 'used up', 'stressed out', 'worn away' and 'unfulfilled'. During the telephone call to celebrate her marathon victory her language had completely changed – she told me she was enjoying meaningful work, she felt on track, confident and even peaceful. She realised her boss wasn't arrogant, just motivated to be successful, her husband's travel was an essential and enjoyable part of his job and she was thrilled to report her parents were thriving in a lovely care home. Different language for feelings associated with different paradigms.

CHAPTER THREE

Negative Nightmares

'We can complain because rose bushes have thorns or rejoice because thorns have roses.'
– Alphonse Karr, French Novelist

I am sure you have had the experience of working with a person you think is negative. Whether the sun has come out after a week of rain, their favourite team won the FA cup final, they achieved glowing feedback in a performance review or a compliment from someone they secretly fancy, they still comment that it is a mistake, will result in something bad happening or shrug it off. They seem to sap the energy in the room, dragging others down with them and are just not good company. But let's remember that different people respond to events in different ways and that the same event can have completely divergent meanings for different people. Events don't have any intrinsic meaning until a human being interprets them. It is that meaning that determines how you feel, whether positive, neutral, negative and a whole range of other emotions.

You may find yourself trapped by worry and negative thinking. One upsetting thought leads to another and another, until there is a whole pile of negative thoughts that you connect together like a never-ending chain. The chain gets heavier with each new thought, dragging you down, sapping your resilience, resulting in worry, anxiety and even fear. Worry hinges on a chain of bad possibilities that can lead to thoughts of catastrophe and 'worst-case scenario' outcomes. *Many of your worries are unfounded.* Negative thinking leads you to believe something is far worse and scary than it really is and on top of that, worrying is about events in the future, something that hasn't even yet happened. When you are trapped in negative thoughts and worry, the chain continues to get heavier and if you are weighed down further you can slip into a state of depression. It's a catch 22 – depression can lead to more negative thoughts and negative thinking doesn't help you deal with your depressive symptoms. Depression literally distorts your perception so that good becomes OK. OK morphs into bad and bad spirals into a crisis. If you only have limited interpretations for why things happen, then thinking positively about change can be challenging to say the least.

Depression acts like a vicious circle because the more depressed you feel the more likely you are to interpret what is happening around you negatively. And the more you frame things negatively the more depressed you will become.

JOHN'S STORY

John ran an IT department in an international bank. He had worked in financial services all of his working life and enjoyed the challenge of keeping up to date with new technology and implementing increasingly secure systems into his environment.

He had weathered the storm of a number of restructurings over the years and given the feedback on each occasion that he was reliable, a safe pair of hands and therefore essential to retain. But each occasion had taken its toll and he had begun to frame that feedback as negative, believing that he had a shelf life and being reliable would not be enough in a world experiencing a technological revolution never known before. In our first conversation several times he repeated his belief that being reliable today really meant becoming irrelevant tomorrow.

His strategy so far had been to keep his skills current, investing his own time and money in learning new applications, attending conferences and special interest groups. But he now wondered if there was any point in so doing as it was inevitable he was for the chop next time round. He went on to lament the prospect of not getting another role anywhere else as he was becoming long in the tooth, outnumbered by bright, flexible, tech-savvy young employees who were cheaper to employ. On top of that, so much more of his work was being outsourced to other countries as a more cost-effective option in providing the bank with the same services a large in-house department could.

He prided himself on being a good manager but worried if he was losing his connection with a new younger workforce. He believed he was being passed over for promotions and cross-functional project opportunities although his own manager assured him that wasn't the case. At the beginning of our time together he was really cross that two colleagues had gained more senior positions yet were less skilled and less up to date in his opinion. He was quite damning of one of them and told me he was worried he wasn't good enough anymore to work such a leading brand.

His manager had suggested a coach as a sounding board for John, after his annual performance review. John was not surprised that the appraisal was not as positive as in previous years. He had had a number of discussions during the year about his emerging negative, pessimistic disposition which he agreed with. But he had no idea how to bust himself out of the pessimism with an impending gloomy employment horizon ahead of him. He contacted me after an argument with his wife who told him he dragged the mood down in the house as all he did at home was complain about work and ruminate about a future of doom.

The brainy bit

John was caught in the trap of negativity, focusing on what was wrong in his life, giving me only negative reflections from his experiences, catastrophising every little setback, blaming himself for everything going wrong and worrying about potential future disasters. The key in his lock was firmly stuck, he was locked into a state of negativity and worry, recognising it, but finding it hard to rise above it.

Alex Korb, a neuroscientist in the department of psychiatry at the University of California, Los Angeles and author of the excellent book *The Upward Spiral: Using Neuroscience to Reverse the Course Of Depression One Small Change at a Time*, suggests a useful difference between worrying and anxiety.[1] He says that worrying is mostly thought based, involving the prefrontal cortex and its interactions with the limbic system, particularly the anterior cingulate, while anxiety involves only the limbic system with interactions predominantly between the amygdala, hippocampus and hypothalamus. He says, 'In essence, worrying is thinking about a potential problem and anxiety is feeling it.'

Korb makes an additional distinction between fear and anxiety – fear is a response to actual, real danger which is happening now whereas anxiety is a concern about an unpredictable event that might happen in the future, that you have no control over – even anticipation of danger can activate the stress response. Anxiety doesn't always have a conscious, thinking element – it can simply be a sensation, like an upset stomach or short-ness of breath. Often there is a physical manifestation of anxiety and you can wind up feeling sick, and you may not recognise anxiety as the cause. My son Gregor was admitted for a suspected appendicitis at the age of eight until a wonderful doctor challenged the cause of his tummy pains and the treatment followed a different route. This had followed the death of his grandfather and a difficult time at school and despite being open and honest in discussions as a family, Gregor had a painful stomach, with bloating for months. He was physically manifesting his mental grief and worries.

The bottom line is that when you are using your pre-frontal cortex to worry, there is no capacity to focus on other areas, like looking ahead to achieve your goals, spending time with family, planning a night out or doing an online grocery shop. Worrying drains cognitive capacity and stops you paying attention to what you are doing. It saps your physical and mental energy making you exhausted, which takes the joy out of your life. All in all, a lose-lose. When John first came to see me, joy was completely absent from his personal dictionary. He even told me that he would find joy in heaven (if of course he got there) but for now he would put up with feeling OK by the end of our coaching engagement.

The upside of worrying

Of course, worrying can be helpful, particularly if the sabre tooth tiger comes for a visit. As discussed in the previous chapter, the reason for the stress response is to keep you safe and alert. And you should be on alert about your performance review if you have been bunking off and going to the pub in the afternoon instead of the office, or worried about your weigh in at your slimming club if you've eaten Creme Eggs all week. Worrying can be put to good use as it causes a debate in your head and gets you thinking about alternatives and consequences. Worrying and negative emotions do have a role. In balance they help you to change and grow. Without them, you may miss opportunities for growth. Sometimes a dose of pessimism or worry is just the thing to motivate you to act, to plan, prepare for the worst, or prevent a nasty outcome.

The problem is that your worry and anxiety circuits may activate or interact with each other too often, keeping you stuck. Fortunately, recognising how your brain works is a key step towards awareness and acceptance which can help to combat the worry and anxiety.

Depending on the situation, the alarm is set off by the anterior cingulate, the amygdala or even the hippocampus. The anterior cingulate controls attention and notices problems. The amygdala is also primed to detect threatening situations while the hippocampus is good at spotting subtle similarities and differences between situations. Learning to use your attention to notice problems, detect the threats and debate alternatives is where a coach can help.

Irrational worries and emotions

Difficulties with emotional regulation can underpin mood and anxiety disorders. Lack of understanding of these emotions or being overwhelmed by negative feelings can result in irrational thoughts that lead you to see threats where there are actually none. And it is a spiral – you worry about having the thoughts and you worry more about not understanding where they are coming from. Sometimes you even consciously know that you are being irrational in your thoughts and subsequent actions and worry further that you can't control it. John became irrationally worried about his one-to-one meetings with his manager, and fixated on potential discussions about his negativity, which took all his mental energy, rather than focusing on meeting his objectives so he could give his manager useful, positive updates. Being in an irrational state can be a horrid, scary place from which to live your life.

In his excellent book, *The New Executive Brain: Frontal Lobes In a Complex World*, neuroscientist Elkhonon Goldberg devotes a whole chapter to the subject of emotions.[2] Neuroscience is now moving beyond the traditional idea that everything is about specific brain regions with particular areas having a dedicated function. For many years popular science told that rational thought processes were controlled exclusively by the cerebral cortex and emotions by the subcortical limbic system in the centre of the brain. Goldberg describes the relationship between the prefrontal cortex and the amygdala – the subcortical structure most often highlighted for emotional control – as a 'two by two circuitry design' in trying to map the functional brain anatomy of emotions – the left and right prefrontal cortex together with the left and right amygdala. In this circuit, he suggests the prefrontal cortex performs an 'editorial' function over the amygdala by 'modulating, modifying and even suppressing the outputs' from the amygdala.

Further work shows that the amygdala is instrumental in mediating a wide range of emotional states – both negative and positive, despite traditionally being associated with only negative emotions, the fear response and memory for unpleasant stimuli. Like the left prefrontal cortex, the left amygdala is more active in response to pleasant stimuli, but its activity is reduced in more negative states. In addition, the ability to actually suppress negative emotions is found to have higher activation in the left prefrontal regions.

Joseph LeDoux is the founder of the Emotional Brain Institute, a Professor at New York University's Department of Psychology and has worked on emotion and memory in the brain for more than twenty years. He points out that the interaction between the limbic system and the prefrontal cortex is a two-way street. In his seminal book, *The Emotional Brain*, he talks about the role of the amygdala and the pathways that evoke a fear response.[3] In a rat it takes almost twice as long for an acoustic sound stimulus to reach the amygdala regions via the cortical pathway than the thalamic route. The faster thalamic pathway provides a warning that something dangerous may be around. By the time your cortex has worked out that a cracked twig underfoot may be a snake, the amygdala is already working on its defence. LeDoux explains that the thalamus is unfiltered and has the role of evoking a response – any response – whereas the cortex has the role of preventing an inappropriate response. The cost of treating a stick like a snake is less than the other way around. When you're threatened, the amygdala is on high alert. When pleasure comes your way, or anticipating pleasure, other structures in the limbic system activate large amounts of the neurotransmitter dopamine. The limbic system sends its stimulation up to the cortex, where these sensations inform our higher order mental structures.

LeDoux explains that connections from the emotional limbic system to the cognitive systems are stronger than connections the other way around. When you are in a spiral of negative, irrational thinking with the chain of worrying thoughts weighing you down, your brain is strengthening the emotional connections and associations making it even more challenging for your prefrontal cortex to put the brakes on. This was evident with John from the outset. His language of 'this always happens to me', 'I never get the recognition I deserve', 'it's always my colleagues who get the interesting projects', using generalisations, only served to strengthen further the emotional anchoring of his experiences.

Negative, irrational thoughts are rarely random. Often, there's a specific underlying emotion, or subject or pattern that triggers your negative thinking and starts the chain linking your thoughts together. Once you become aware of your personal patterns or themes, it becomes easier to break the links and sever the chain. Through our work, John identified that he was jealous of the success of his friends and colleagues, an emotion that he found repulsive, so even the act of identifying and surfacing it, was hard work for him. He was triggered into negativity when he witnessed success of others around him and he felt jealousy at a deep emotional level. He didn't feel good enough to compete and his jealous emotions caused him to spiral further into self-blame, pity and the view that he didn't deserve success anyway.

Ruminations

Worries in daily life often take the form of subconscious ruminations over possible threats to your well-being. Repetitive negative thoughts of, for instance, anger, guilt, regret, envy, or blame are known as ruminations and have been well researched both in the neuroscience and psychology worlds. A tool used in cognitive therapy, termed the ABC Technique of Irrational Beliefs, by psychologist Albert Ellis can help you understand what is happening that keeps you worried.[4] The idea is that external (or activating) events (A) do not cause emotions (or consequences) (C) but are caused by beliefs (B), in particular, irrational ones. The worries that plague us unnecessarily are the result of irrational beliefs triggered by the event. It is the belief, not the event that leads to the emotional consequence. For example, you may worry that you won't like your colleagues in a new job you are starting and therefore should turn down the offer. This worry comes from the belief that as an introvert, you're not a team player and therefore you won't contribute to the team performance and they won't like you. The triggering event could be a comment made by your new boss that the team are excited to meet you, or you overhearing a discussion about the fun the team all had on a regular night out. It's the belief, the irrational one, which causes you to experience unnecessary worry and anxiety.

Neuroscientific research on rumination and repetitive thinking helps us understand the brain mechanics of dwelling on negative thoughts. A 2015 Stanford and Roosevelt University study in a meta-analysis of previous research identified that depressive ruminations are more likely to emerge when increased blood flow to the subgenual prefrontal cortex synchronises with the default mode network.[5] This is a network of brain regions that are active when your mind wanders and you become lost in nostalgic thoughts or daydreams. On an EEG, the brain typically appears to be in a wakeful state of rest when the default mode network is activated. Normally the subgenual prefrontal cortex helps to balance your reflective process supported by the default mode network so you can consider problems in order to develop solutions.

The researchers believe that increased connectivity between the subgenual prefrontal cortex and the default mode network can backfire by creating a vicious cycle of rumination in people who are experiencing negativity and that depression distorts a natural process. So, there are new neural networks linked to rumination. However, in depression it seems that the subgenual prefrontal cortex runs riot, stopping normal self-reflection in its wake. This may be one reason that electrical stimulation of the subgenual prefrontal cortex is helpful for some patients with severe or treatment-resistant symptoms of depression.

From a positive psychology perspective, there are infinite benefits to breaking free from ruminative thinking such as becoming more creative and open to possibilities. In some ways, rumination is the opposite of focusing on positive emotions. A possible way to dampen the ruminative neural pathways may be finding daily positive routines such as exercising, affirmations, visualising success to activate the ventral striatum, moving you towards pleasure.

It's a catastrophe

Once started, it is so easy for a negative thought to spiral into a potential disaster. You worry more about an impending catastrophe by envisioning the worst possible scenario. Habitual worriers like John believe that thinking and anticipating all the possible negative outcomes helps them feel safer and more prepared if the depicted scenarios become real.

Psychologists describe catastrophising as a cognitive distortion where you tend to predict the worst-case scenario and leap to the conclusion that a catastrophe will result. Sometimes this stems from an initially reasonable worry and then through a series of assumptions underpinned by irrational beliefs your worry snowballs out of control. Your headache becomes a fatal brain tumour in your mind and you spend all your time on Google trying to confirm it. Not getting a first-class degree means you will never have the career you dreamed of. Before you know it a situation you are concerned about explodes into a full-blown catastrophe.

A teacher friend recently told me a story that a mother had insisted on accompanying the group on a school trip into central London. Her rationale was that due to the increasing threat alert level in the city, she wanted to protect her 15-year-old daughter on the 'dangerous' London underground. She evidently became fixated on the journey into London that day, despite the fact that her daughter travelled into school every day for five stops on the underground network, and that her presence wouldn't have changed the outcome of any terror activity. My friend told me she became genuinely irrational and almost hysterical in her reaction when told that the school didn't want to set a precedent by inviting parents along, particularly for that age group.

University College London and Cambridge University researchers have identified the habenula as the part of your brain responsible for predicting negative events, using electric shock or monetary reward cues.[6] They also believe this pea-sized cell mass plays a role in your ability to learn from bad experiences. It tracks your experiences, responding more the worse something is expected to be. The habenula has previously been linked to depression, and this study shows how it could also be involved in causing symptoms such as low motivation, pessimism and a focus on negative experiences. A hyperactive habenula could cause people to make disproportionately negative predictions and imagine worse case scenarios.

A genetic predisposition to negativity

The Universities of British Columbia, Cornell and Toronto, together with the Rotman Research Institute and the Centre for Addiction and Mental Health in Toronto, collaborated in research that found that a genetic variation can significantly affect how people see and experience the world and that biological variations at the genetic level can play a significant role in individual differences in perception.[7] The ADRA2b deletion gene variant can cause people to perceive emotional events, particularly negative ones, more vividly than others. This gene influences the hormone and neurotransmitter noradrenaline. Previously found to play a role in the formation of emotional memories, the new study shows that the ADRA2b deletion variant also plays a role in real-time perception.

Participants with the ADRA2b gene variant were more likely to perceive negative words than those without the gene while both groups perceived positive words better than neutral words to an equal degree. The researchers say they may be more likely to notice hazards, seeing areas they may potentially slip rather than noticing the beautiful surroundings, or see angry faces in a crowd of people. The gene variant may be helpful in some circumstances but by seeing negative aspects first, may trigger a spiral of worrying in others.

Uncertainty makes you anxious

The one thing about uncertainty is that you can be certain it exists in your life. You do not live in a predictable and stable environment – economic, political, and cultural change surround you at the macro level as well as all the smaller but continual daily uncertainties all around.

Like John, worriers cope by avoiding uncertain situations to try to keep things under control as much as possible. A couple of sessions into our coaching, John admitted that he had cancelled his most recent one-to-one catch-up session with his manager, fabricating an urgent meeting in the business. He also had taken to going home a little later, hoping to avoid the 'how was your day?' chat with his wife if she was busy with the children. People often associate uncertainty with a potential danger and a negative outcome, despite the fact that such outcome might not even occur in the first place. John's one-to-one meeting may have been positive, motivating and fruitful and his wife may have shared some happy news, but John chose to avoid the possibility of 'bad' conversations. We discussed that it's not bad, it's just unknown and that uncertainty is a certainty in life. People all react differently to uncertainty but it can make you feel uneasy or unhinged, trapping you in negative thoughts and worry and can result in avoidance.

Recent research from University College London shows that uncertainty is even more stressful than knowing something bad is definitely going to happen.[8] Participants played a computer game, where snakes were hiding under certain rocks. They learned over time which rocks were more likely to hide a snake and if a snake appeared, they were given a painful shock to their hands. The game was designed with fluctuating and high uncertainty, tracked by a complex computer model which also estimated levels of participants' uncertainty for each choice made, using the guesses they made over time. Their stress levels were also measured physiologically, by pupil size and perspiration.

The study found that players were most stressed when they were more uncertain about the situation than when they were certain about either getting or not getting a shock. It seems that people feel better about knowing what's coming – even if it's painful – than not knowing. When predictability was at 50%, when people had absolutely no clue whether they were about to get shocked, stress peaked.

So, it is the wondering whether something is going to happen that causes the worry. Wondering if you are going to miss your train rather than knowing you left the house too late to catch it. Or not being sure your partner is cheating rather than knowing they are. Or, in John's case, wondering if his manager had asked for a meeting for a project update rather than knowing that he was going to get feedback on his negative disposition. The wondering and uncertainty cause higher worry and stress levels.

You know from the first chapter that the striatum moves you towards rewards and positive outcomes and away from negative punishing ones. Dopamine floods the striatum just as much whether good news or bad news is coming your way. And when uncertainty of a good or bad outcome is high, stress is guaranteed. In trying to trigger some corrective action, it activates your sympathetic nervous system, opening up your sweat glands, dilating your pupils and energising your muscles ready for action.

The snake study suggests that strategies and gadgets that give us the ability to predict outcomes may be quite calming – like a notification about how far off the next motorway exit is, seeing your Uber driver's journey en route to collect you, and displays or apps showing the arrival time of the next bus.

It is easier to plan for the future when you know what the deal is, even if it's bad. When applying for a job you'll probably feel more relaxed if you think it's a long shot. My elder daughter was so unsure how she had done in her GCSEs that she had negotiated six possible A level options with her teachers. She wanted a specific combination and frankly I can only describe it as a summer of stress for the whole family, there were snakes and electric shocks under every rock. She proved to me that the most stressful scenario is when you don't know what to expect. It's the uncertainty that makes you anxious.

Pessimism can result in doing nothing

Optimism and pessimism can most simply be described as expecting a positive or negative future. They are generalised modes of thinking that vary from person to person depending on the context. You can be optimistic about one area of your life such as having a successful career, but pessimistic in another such as never being able to find a life partner. It is normal to experience happier and deflated days where you see the world differently.

But some people have a consistent tendency to think, feel and behave in an unbalanced way, more generally as optimists and pessimists across most aspects of their lives. An optimistic person views the world as full of potential opportunities. The pessimist sees mainly the negative aspects of everything around and is likely to have little hope for the future. The optimal balance for most of us is realism which may be described as cautious optimism. Being overly optimistic may result in recklessness or negligence such as avoiding contraception or not preparing for a job interview. Being overly pessimistic can lead to apathy and demotivation when faced with a challenge, believing that there is no point in trying. This is where John was at with his latest project, saying it was unlikely to work out well anyway, so why put in the extra mile? Pessimism can also spiral into low mood, anxiety and depression when your life becomes utterly miserable.

A literature review from the Institute of Cognitive Neuroscience, University College London on the neural basis of optimism and pessimism suggests that they are associated with the two cerebral hemispheres.[9] High self-esteem, a cheery outlook veering to the positive and an optimistic belief in a bright future are associated with activity in the left hemisphere. On the other hand, a sombre perspective and a tendency to focus on negative elements, low self-esteem as well as a pessimistic view on what the future holds are linked more to right hemisphere activity. The researchers summarised that this hemispheric asymmetry in mediating optimistic and pessimistic outlooks results from various biological and functional differences between the two hemispheres. They say that '… the right hemisphere mediation of a watchful and inhibitive mode weaves a sense of insecurity that generates and supports pessimistic thought patterns. Conversely, the left hemisphere mediation of an active mode and the positive feedback it receives through its motor dexterity breed a sense of confidence in one's ability to manage life's challenges, and optimism about the future'.

Reducing negativity

In recent years, there has been a huge interest in the discussion around happiness and positive thinking. Emanating from the realm of positive psychology all around us there is an explosion of books, blogs, studies, posters, journals, podcasts, sticky notes, internet tips, and daily social media motivation quotes to name a few. The positive psychology ethos suggests that getting rid of sickness, disability, depression, crime and the other problems in life is important, but not enough. People should be able to not just survive life, but to thrive and enjoy it. The typical approach for dealing with mental illness is to wait until a person shows signs of disorder, then provide treatment. This is like taking your car to a garage when it stops working. Just as some of your car problems can be avoided with regular services, rather than waiting for it to break down and use your AA membership, positive psychology suggests that by proactively taking care of your mental health, mental illness can be prevented, or at least be less severe.

It is fantastic that the mental health agenda has come to the fore across all walks of life. It is helping people to understand that the brain gets sick and just like you wouldn't tell someone to snap out of it if they had a broken toe, there is now more recognition that you shouldn't tell them to do so if they are anxious, worried or depressed. This heightened awareness is helping people to change the narrative about their concerns, giving themselves permission to talk more freely about their feelings. This all comes with a health warning. Over focus on positive thinking can lead to excessive optimism, setting unrealistic expectations, being unprepared for failure or down times. It may affect your motivation to consciously look after your health as you believe you will always be well. It can even make you come across as insensitive if you are trying to cheer up someone who is genuinely sad. They may need you to listen and acknowledge their feelings, not chivvy them out of it. Checking in with yourself regularly to see if you need a hit of realism is no bad strategy.

Managing your emotions

As well as progress with attitudes towards mental health, we are also experiencing an acceptance of acknowledging and being open to the emotional side of your personality. Daniel Goleman spearheaded the emotional intelligence revolution which is now firmly on the agenda of most organisations.[10] Many of the MyBrain corporate clients and those of our Practitioners have development programmes and competency frameworks that include learning about the importance of emotions, being open to expressing feelings and connecting with others.

Like John, if you are caught in a spiral of negativity that you recognise isn't serving you well, the good news is that you can override your limbic system's tendency to let your emotions control your life, but it takes mental effort and a desire to do so. You have to decide to be the one in charge of your emotions, or your emotions will take charge of you. This is an important point as we all know, unless we want to do something, it is unlikely to happen – just look at all the repeat business Weight Watchers gets. Their purpose is fantastic, but many people lose motivation when they leave. For a discussion around discovering your purpose, see Chapter 6 on motivation.

Susan David's excellent book *Emotional Agility* is an advert for positive psychology, discussing the power of happiness in our daily lives, work and relationships.[11] She argues that the way we perceive our inner selves through our thoughts, feelings and narratives is the key determinant to how we live and the success we achieve. Keeping a negative self-image, like John, is destructive, and can erode your potential for success.

She discusses adaptation as the key to transforming yourself to achieve the success and happiness you want.

Facing your thoughts or feelings is arguably the most difficult thing to do but is also the most necessary to make any change. She suggests a strategy of curiosity, accepting both the difficult and positive thoughts equally in order to see them for what they are. Just as a seed cannot grow from concrete but a mix of soil, water and sun, in the same way you must allow a combination of experiences and thoughts to shape you, which ebb and flow as life progresses, rather than holding on to beliefs that don't serve you well. To become emotionally agile, you need to allow flexibility into your life and allow your beliefs to change. You cannot expect the same rules or actions to apply across all the different experiences and circumstances that you encounter in your life and as your context changes. Emotional agility is an essential, self-serving life skill.

Professor of Neuroscience at the University of California, Los Angeles, Antonio Damasio, in an interview with the MIT Technology review said that what is distinctive about the human species is that you connect the fundamental processes of life regulation such as emotions and feelings with intellectual processes to create a whole new world around you.[12] Emotions instruct you about how to act, whether to approach or to avoid, whether to fight or to run, or whether good or bad may result from them.

Learning to change your thought patterns

A coach can help clients learn to deal with their emotions by changing their thoughts and actions to be consistent with new thoughts. Coaching can retrain the rational prefrontal cortex to take control over the irrational emotions of the limbic system and listen to useful ones. Your prefrontal cortex receives information from the outside environment through your senses, which then shoot messages through some of the limbic system enabling you to feel before you think. However, if you speed up or strengthen your cortex's natural ability to inhibit the limbic system, you can change your feelings before they have a chance to impair your behaviour or judgement. In this way, the coach can serve as your 'cortex' when your own is overridden by negative emotions that are ruling your limbic system and show you the way to take control.

Changing your thoughts is not the same as suppressing emotions. It is important to acknowledge, discuss, label and detach from them. The late Daniel Wegner was a psychologist known for his work around thought suppression with people unable to stop thinking about a white bear.[13] He found that there can be significant consequences when you try to push away thoughts and feelings calling them the rebound effect. These strategies can backfire and actually result in an increase of the intensity of the thoughts and emotions that are being suppressed.

In a review of studies, a team of Brazilian neuroscientists showed that cognitive-behavioural therapy interventions change the neural circuits involved in the regulation of anxiety.[14] There seems to be beneficial effects of therapy and other change-focused work on the neurobiology of negative emotions. After you learn to change the way your limbic system reacts, you have less reason to worry because you're not processing your experiences the same way anymore.

It is essential to spend time talking through your feelings and identifying your irrational beliefs. Many of those beliefs lead you to see threat where there isn't any and sometimes involves your perceived need to live up to other people's expectations. A strategy is to remind yourself of possible good outcomes to reduce the negative impact and then think through a plan to deal with the outcome. Planning your response to stressful situations can increase noradrenaline and calm the limbic system. Exploration can help you discover a more realistic balance for yourself by setting goals, milestones and expectations that work for you. In our work, I encouraged John to borrow my belief in him, that he was more than good enough as a way to show him the means to take control and pave the way for exploration.

Reframing

Learning to acknowledge, use or counter your illogical thoughts may involve substituting your negative perspective with more neutral or positive thoughts. Reframing is an important coaching tool that can help you to view your situation through a different lens and so develop different thoughts and feelings about it. It uses the metaphor of a picture frame to help a client explore possibilities. If your favourite photograph was framed in a beautiful, sophisticated solid silver frame and you changed it to reside in a boxy frame bordered in black shells, it would change how you viewed the picture. Similarly, if you had a frame with your picture zoomed in or zoomed out of the familiar perspective, you would view it in a different way. Same content, different context, different perspective.

Reframe not getting a promotion as an opportunity to look at a different company or department, your sprained ankle as a chance to rest, the coffee shop being closed as an opportunity to save money or the intern being dumped on you as a way to reduce your workload.

Reframing is the first part of what I call the three Rs. Start with reframing to help identify areas for refocus before moving onto re-evaluating next steps for these new focus areas. Reframe, refocus, re-evaluate. In our second session, acting as a devil's advocate, I taught John the reframing technique. I reminded him that the meaning we assign to events is through our own lenses on the world – and importantly, how that meaning may not be the actual truth. We played a metaphorical game of ping pong, bouncing back and forward ideas for what his colleagues' promotions could mean and why they had happened for them and not him. As a result, he was able to refocus and re-evaluate what was important to him in a job and to start to plan towards proactively looking for a new opportunity with his manager's support. It took effort and courage to have the conversation with his manager but he felt encouraged when his manager's response was positive. In the same conversation he asked the manager what he meant in his feedback that John was reliable. His manager instantly responded that he meant that John was an essential member of the team, he was performing a critical role well and being reliable meant he didn't have to micromanage him. What a reframe that was for John. One word – reliable – and two completely different interpretations. His manager meant he was essential to the team and totally trustworthy, whereas John had interpreted it as being potentially irrelevant and mediocre.

Labelling

Labelling helps you become detached from your inner chat, thoughts and feelings so you can see that they are just emotions, not you personally. These emotions are not bound to your body forever and are not an essential part of your survival in future situations. Becoming detached from them will enable you to talk through them in a less defensive and more honest way. This won't happen overnight but with effort through discussion and exploration it is a key step in you feeling far more in control over your actions and decisions.

Distortion labels

Negative thinking is often a result of our minds convincing us that something is true – when it isn't. You can develop some faulty and unhelpful distortions over time which become patterns of thinking or believing that are false and have the potential to cause damage. Some common ones include:

- Generalisations – you believe because something bad happened once it will always happen, or conversely something will never happen if it hasn't happened before. I can still hear my teenage daughter slamming the door after shouting 'you will never allow me to go out until midnight'. John and I kept a tab of how many times each session he used the words always or never. Just by catching him doing it and pointing it out – fourteen times in our first session – enabled him to start noticing his language himself. By generalising so frequently, he was continually convincing himself that he would 'never be good enough' and would 'always lose out to his colleagues'.

- Jumping to conclusions – your boyfriend hasn't been in touch, so he is being unfaithful.

- Personalisation – your colleague chose a particular project from a list because she knew you wanted to do it.

- Blaming – it is never your fault, or it is always your fault.

- Heaven's reward fallacy – struggle, suffering and hard work will result in a just reward, like going to Heaven. If the reward doesn't happen, it can result in anger, bitterness and even depression.

Before you can challenge and work to combat distortions you must of course identify them, they are often operating at a subconscious level and you will have solid neural pathways for your favoured distortions. They are not a life sentence and once surfaced, it can be liberating to find another path.

Take control by committing

Planning and committing to focus on one change at a time from your negative arsenal, one goal or accomplishment, gives no space for confusion or overwhelm and can give you back some control. When you achieve a small win, you will slowly become motivated to do more and may face your negative habits with more gusto and strength. Taking a first step is something to be proud of and reframing your actions to pride, courage and commitment gives a more positive platform for you to assess your progress. The ethos of self-efficacy helps you to gain control over your negative behaviour when you see yourself being able to exert that control, act and achieve a result. As you gain more confidence from improved decisions, from overcoming your negative self-chatter, you will gradually find that your attitude changes and you will be less dominated by negativity and pessimism. John committed to me that he would never again fabricate another meeting to avoid his manager. He committed to a monthly one-to-one calendarised schedule of meetings with his manager making sure there was time to include a discussion and feedback around any negative behaviour his manager perceived.

Remember, when you commit to something it may be a turbulent ride of emotions just like learning to snowboard as an adult. You fall, get up, become bruised, develop fatigue in muscles you haven't felt for years. Being a beginner on the slopes can be exhausting but what hurts more is making the choice not to act by continuing to indulge in negative emotions and never knowing the achievement or liberation of pulling yourself out of your spiral. And this is an important point. Not doing anything or avoiding action is a choice. When John realised this, it provided him with more motivation to change his negative spiral – after all, he could choose to do so. The more he practised the snowboarding of change, the easier it became as new neural circuits were laid down.

Calm the catastrophising

As well as the distortion effect of catastrophic thinking, it can also trigger the actual outcome of the catastrophe. If you turn your attention to something, switch on your internal monologue to that catastrophe, you are almost increasing your chances of a self-fulfilling prophesy. Worrying that you haven't done enough revision for your history exam and feeling anxious before walking into the exam room will inevitably spike your cortisol level, reducing your ability to react effectively. If you have also told yourself that failing will be a disaster and your teacher Mrs Smith will be angry, then you see Mrs Smith as the invigilator, then guess what? You are heightening your stress even more. A further layer of panic in the moment leads you to feel physically sick and then tell yourself you deserve to feel this way as you didn't revise enough and you are a worthless loser. On and on it goes.

Your mind is effective at creating catastrophic thoughts and you are good at convincing yourself that your thoughts are true. Fortunately, there are strategies you can try to calm your anxiety and empower yourself to act.

The trick is to notice your thoughts. Catch yourself when your thoughts move from an understandable worry to an unlikely scenario. Pay attention to patterns. 'That's interesting, whenever Mrs Smith is around, I get nervous that I haven't done enough work. I don't think this with other subjects and I do the same amount of revision. What is it about history that I'm concerned about?'

When the volume turns up and you move to catastrophic thoughts try countering this with reassuring thoughts such as, 'Wow, I sure am over reacting. I don't deserve to feel this way as I have done some revision.

I am beating myself up far too much, I will pass and go into the exam and do my best.'

Noticing is a technique that requires you to take a breath, pause and create a tiny space to catch yourself, reboot and consciously look the catastrophic thinking in the eye. It is in this pause where you can engage your prefrontal cortex as your friend and strengthen the neural circuits from it to your amygdala, so calming your emotions.

Once you identify the catastrophic thinking, grab some control back and take some action. There is nothing like doing something proactive to make you feel more positive. For exams you need to feel more prepared or practice mindfulness techniques to calm your amygdala in moments of nervousness. If you fear your teenager is going to stay out all night, have an open adult-to-adult discussion to discuss boundaries. Face your fears – at least if you are open to solutions and her ideas, you will have a different frame to work with, rather than worrying she will be out all night, ruminating and continuing to be stuck with an angry teenager. For John, he threw his hat in the ring immediately the next high-profile project opportunity presented itself – and he was delighted to report he became the next project lead.

Dealing with uncertainty

The ultimate reframe is to accept that uncertainty is unavoidable as it is all around you. In any situation it helps if you can identify what you can influence rather than what you have no control over. One thing you can change is your response to uncertainty, and to identify your unhelpful responses to an uncertain stimulus rearing its ugly head. It is important not to use avoidance tactics as a coping mechanism. This can rapidly become a habit and can result in decreased confidence making change even harder. Make the choice to act, not avoid.

Use the noticing technique as described above to vocalise your feelings about the uncertain situation. In addition, ask yourself what a good outcome would be as well as your usual mantra of a bad outcome.

'So, another exam, this time history. In reality I'm rarely sure of a great result in this one, but I have never actually failed. So, if I give it my best shot I could nail it.'

Challenge your habitual response to uncertainty, identify other more useful responses and commit to making one change at a time.

Remember the snake experiment earlier in this chapter?[15] The researchers reported a secondary fascinating finding that the participants whose stress response mirrored actual (not imaginary) levels of uncertainty performed best on the task. In other words, their sensitivity to uncertainty gave them an edge when it came to predict which rocks to avoid, even though they couldn't avoid shocks in the long run. So uncertainty may be motivational for some people (see Chapter 6 for more on motivation).

Gratitude and thankfulness

No matter what the presenting issue is, I always encourage my clients to commit to journaling their gratitude. Gratitude can be a life saver for negative feelings because it doesn't necessarily have to be about your personal woes and limiting beliefs. You can be grateful for sunshine in the sky yet not have two pennies to rub together, or grateful you have an income even if you hate your boss. You may be negative about not having much money or having a poor relationship with your manager, but it doesn't stop you stepping out of your situation to focus your attention on areas that could bring you joy.

Practising gratitude is something I have done with my family for years. On a Sunday evening at dinner, round the table we share at least one experience or thought from the previous week we are thankful for. It can be anything, even seemingly materialistic 'first world' things, like Sky Plus or working Wi-Fi, or comforts like a warm bed or a cup of tea, or a way of expressing love like being with family or (I even heard from one of my daughters – once) 'I am grateful for my brother'. Gratitude improves mood, calms worrying and increases feelings of social support. It improves dopamine activity, boosts serotonin, enables improved sleep and physical health in other areas. A detailed discussion on gratitude is found in the next chapter on loneliness.

Comparing to others

One of John's deep-rooted emotions surfaced through the coaching conversation was envy. He found this challenging to discuss as it flew in the face of his core values and initially he became even more upset with the character that emerged. When he started his gratitude journal it was clear he was in the habit of comparing himself to others. Further work revealed that he lived in a world where he perpetually measured himself against others and assessed his worth on multiple comparisons against people he both liked and disliked.

Realising that other people are worse off than you is not the essence of gratitude. Comparisons with others less fortunate than you does not have the same benefits as being thankful for what you have. Gratitude requires an appreciation of the positive aspects of your own situation. Sometimes noticing what other people don't have can help you see what you can be grateful for, but you have to turn it around and show appreciation for what you do actually have, for it to have an effect. It does not matter what other people do or don't have.

The gratitude journal was a huge catalyst for John to move forward and make positive changes in his life. Towards the end of our formal coaching time together, he wrote a gratitude letter to his former self, detailing why he was thankful for his experience of being stuck, for entering the world of envy so he could look it in the eye and renounce it wholeheartedly. He recognised it was holding him back and he didn't need it anymore. It was a powerful experience for him to share and one he says he still believes was a life-changing turning point.

Turn up the optimism

Simply imagining the possibility of a positive future can help to strengthen the brain circuits involved in optimistic thoughts. It's like shining light into your dark spaces. You don't need a deep-seated belief that they will happen, just visualise that possibility. It is possible you will finish the 10k in a sub one-hour time, it is possible you will be invited back for a second interview, it is possible you won't get overdrawn this month. These thoughts activate the ventral anterior cingulate which helps to regulate the amygdala, therefore managing the brain's negative bias. Taking it one step further and expecting something positive to definitely happen also activates the ventral anterior cingulate as well as the prefrontal cortex. Here, the rational cortex is turned on to communicate with the amygdala to help keep it under control.

The language of change

The left hemisphere governs the rules and habits we operate by including the rules for language. Using language tends to boost brain activity on the left side. Writing things down in your journal or saying things out loud can also help to stop rumination and going over and over things in your mind which is a right hemisphere function. When it comes to gratitude, visualising that positive outcomes are a possibility or a definite, pausing to notice when you are catastrophising, in an uncertain situation or comparing yourself unnecessarily to others, the benefits of writing down your thoughts, like John did, are enormously helpful for changing your neurology.

CHAPTER FOUR

Loneliness Hurts

'The most terrible poverty is loneliness, and the
feeling of being unloved.'
– Mother Theresa, Missionary

Humans are social mammals and were not designed
to be solitary creatures. We evolved to survive in
tribes and today we organise ourselves into communities. The need to interact is deeply ingrained in our
genetic code. So much so that the absence of social
connection triggers the same, innate instincts as hunger,
thirst, physical pain and overall survival. Social pain,
also known as loneliness, evolved because it protected
the individual from the danger of remaining isolated.
Our ancestors depended on social bonds for safety and
for reproduction to pass on their genes through their
offspring. Feelings of loneliness told them when those
protective bonds were endangered or deficient.

Loneliness is not necessarily about being alone. Instead it is the perception of being alone and isolated. It's perceived social isolation, or the discrepancy between what you want from your social relationships and your perception of those relationships. It is a common human emotion, complex and unique to every individual. If you feel alone and isolated, then that is your definition of loneliness. It is a common misconception that just living alone or being a quiet introvert makes you more susceptible to loneliness. That couldn't be further from the truth and it is wrong to generalise. Many people who consider themselves extroverts can feel lonely and many people suffer from loneliness despite living with a large family or being surrounded by friendly colleagues. Not everyone who lives alone is lonely and not everyone who is lonely lives alone. Loneliness is a human condition. That means it affects all humans, including introverts, extroverts, and everyone in between.

Loneliness is fast becoming one of the biggest health challenges of our times, and it's not just the aging population that is fuelling that. A study by the British Red Cross and the Co-op shows that over 9 million people in the UK across all adult ages – more than the population of London – are either always or often lonely.[1] Loneliness is a serious health issue and building awareness and tackling it will also likely have a positive health benefit, taking the strain off the NHS and social care services.

Family and friends aren't as closely bound as in the past since the industrial revolution opened up the world. We no longer live in the same village for generations and often don't have the same family connections. The social infrastructure has changed and relationships are formed and replaced more easily today. And then there is the proliferation of technological innovation. We are going through times of profound social change, with technology enabling us to remain in touch with others without actually having to hear a voice or see a face. I remember the early days of social media with Friends Reunited, the business portal LinkedIn and the capability of Skype to enable video of what was previously a telephone conversation. At the time, I was in awe of how brilliant these new technologies were in expanding our connections and adding to our communications reach. Technological capability has exploded exponentially with Facebook, Twitter, Instagram, Snapchat, Tinder, YouTube and numerous other apps and programmes you can join and find friendships, life partners, connections and opportunities that just didn't exist in the past. In the last twenty years, many of our face-to-face connections have been replaced with social networking. Social media can be a great way to promote face-to-face communication, however if used as a replacement it certainly can increase lonely feelings.

Despite this unprecedented global human connection, more people than ever are suffering from severe loneliness. A study from the Office for National Statistics using the Community Life Survey in England from 2016 to 2017, from more than 10,000 adults, found that about one in twenty people always or often felt lonely.[2] Although there has been much focus on the isolation of elderly people, this study found that young adults are more likely to feel lonely than older age groups, women reported feeling more lonely than men, renters more than home owners and people who felt that they belonged less strongly to their neighbourhood reported feeling lonely more often.

In the workplace loneliness is a topic of concern to organisations and can have a significant influence on employee well-being, motivation, work performance and team effectiveness.

DAVID'S STORY

I was recommended to David by one of my therapist friends. He was clear he wanted to work alongside a business coach to look forward and regain his joy for work and his colleagues. He was UK Marketing Director of a leading global software player, an exciting organisation to work for with regular mergers, acquisitions and divestments. He had worked for the business for twenty years from its early start-up days through its meteoric rise to a household name, had relocated to Singapore and Hong Kong for a number of years and had a track record of international marketing success.

I was immediately struck by David's natural warmth and interest in others. When we met initially he said he wanted to make me aware of his personal circumstances so I would have a full picture of his 'back story' as he called it. He had two grown-up sons and a two granddaughters. He had been divorced for fifteen years – an event he described as positive for all the family and an honest congruent decision made by the couple. He and his wife had simply grown apart, their life goals became misaligned but they had strong mutual respect and remained firm friends, continuing to share the care of their boys. They had both subsequently dabbled with romance but neither had remarried. Eighteen months ago, his ex-wife had died after a year-long cancer battle and David had moved her to his home so he could be with her in her final months. He recognised this may have been viewed as unusual but he felt it was the best gift he could give to a woman he had such respect and love for. He had taken a six-month leave of absence from work to help care for her and on his return he said he just didn't feel he fitted in or belonged any more.

Before his time out of the business, he was part of a team tasked with the most radical restructure he had experienced across the sales organisation integrating two acquired organisations. Both new companies had a much younger workforce, many millennials from high-tech, digital marketing backgrounds used to constant innovation and experimentation. On his return to work he found the restructure had been more extreme than expected and while he had effectively been promoted with a larger team and wider reach, not one of his original team members now reported to him. In addition, four of his peer group, his 'cronies' as he called them, had moved to different roles, three on overseas assignments. It was a completely different place to work.

Before engaging a coach, David had seen my therapist friend, wanting to be sure his feelings were about changes at work and not caused by emotions relating to the loss of an important relationship in his life. He was able to articulate his emotions clearly, acknowledge his grief, the impact of his recent personal experiences on both his current home and work life. He felt he had unpicked his feelings in a detailed, helpful way with his therapist and was now clear he wanted to look to the future with support from a business coach to help him deal with his unhinged feelings of not belonging at work. It took a couple of sessions before David was able to fully define his coaching goals of gaining strategies to deal with his loneliness at work and indeed becoming comfortable using the word lonely when referring to himself.

David assured me he was happy in his personal life. He was a member of a golf club, playing every weekend, had wide interests and hobbies, was fit, healthy and active, enjoying regular time with his family and friends.

The brainy bit

Loneliness is a biological mechanism that pushes people to find the social interaction that they need and may lack. We need other people as our prehistoric ancestors required company in order to survive, the presence of other human beings ensured protection and support, both for themselves and for their offspring. The need to belong, for meaningful social connection, and the pain we feel without it, are defining characteristics of our species. Our brains still think we need to be surrounded by others to survive and thrive. Societies are built on generations of complex social interactions and our social skills beyond survival are deeply rooted in our neural circuitry.

A chronic state of social isolation is terrible for your mental well-being increasing your chances of developing anxiety and depression. Loneliness becomes an issue of serious concern when it settles in long enough to create a persistent, self-reinforcing loop of negative thoughts, sensations, and behaviours. It's also a bigger problem than a mental health experience manifesting in physical symptoms and consequences. Research shows lacking social connections is a comparable risk factor for early death as smoking fifteen cigarettes a day and is worse for us than well known risk factors such as obesity and physical inactivity.[3] Lonely people are more likely to suffer from heart disease and stroke,[4] and overall, loneliness increases the likelihood of mortality by 26%.[5]

Lonely brains are different

Research on the effect of loneliness and the brain suggests that social isolation promotes increased vigilance
for social, in contrast to non-social, threats. University
of Chicago researchers showed that lonely people
are more alert to threats and the possible danger that
strangers may pose, because their brains become more
active in a social situation.[6] When you feel socially
isolated your nervous system automatically switches
into self-preservation mode, making you more aggres-
sive and defensive – even if there's actually no threat.
In an experiment where loneliness was defined as a
subjective feeling of isolation, rather than as numbers
of friends or close relatives, brain imaging techniques
were used when subjects engaged in a Stroop test – a
test of cognitive functioning.[7]

The researchers found lonely people became highly
vigilant when the words were regarded as socially
negative (such as solitary or alone), where non-lonely
people responded in similar ways to both social and
non-social negative words (such as hostile or vomit
respectively). The visual cortex becomes more active
while the area in the brain responsible for empathy
becomes less active, concluding that lonely people's
brains are conditioned to tune into social threats
faster than what is considered normal. In addition, the
research suggested that lonely people subconsciously
are alert for negative possibilities.

A study from the University of Chicago and Dartmouth College used pictures of people versus objects and fMRI scanners to show neural mechanisms differentiating social perception in lonely and non-lonely adults.[8] Lonely people showed decreased activation of the ventral striatum to pleasant people than pleasant objects, while non-lonely individuals showed the reverse suggesting lonely people are less rewarded by social stimuli. Increased activation in the visual cortex was also observed when lonely participants viewed negative, unpleasant social scenes of people rather than of objects, suggesting that their attention was drawn to the distress of others. For the non-lonely group, participants showed greater activation of the right and left temporo parietal junction to pictures of people rather than objects. The researchers say this fits with the notion that non-lonely people are more likely to reflect on the perspective of distressed people they see.

In the book *Loneliness: Human Nature and the Need For Social Connection*, a founder of the field of social neuroscience John Cacioppo and science writer William Patrick trace the evolution of these forces, showing how, for our primitive ancestors, survival depended not on greater brawn but on greater commitments to and from one another.[9] They discuss that the pain of loneliness engendered a fear response so powerfully disruptive that even now, millions of years later, a persistent sense of rejection or isolation can impair DNA transcription in your immune cells. This disruption impedes thinking, will power, and perseverance, as well as your ability to read social signals, exercise social skills and regulate your emotions. All of these impairments combine to potentially trap you in self-defeating behaviours that reinforce the isolation and rejection that you fear.

Loneliness, rejection, and pain

John Bowlby, the developmental psychologist who pioneered Attachment Theory, wrote in his trilogy *Attachment and Loss*: 'To be isolated from your band, and, especially when young, to be isolated from your particular caretaker is fraught with the greatest danger. Can we wonder then that each animal is equipped with an instinctive disposition to avoid isolation and to maintain proximity?'[10]

Social isolation research in the Social and Affective Neuroscience Laboratory at the University of California, Los Angeles used a computer game called Cyberball, where players toss a virtual ball back and forth to each other.[11] At a certain point, the players deliberately stop tossing the ball to one player. Scans of people excluded from the game showed that brain activity was similar to what is observed when someone is in physical pain.

Brain scans of the test subjects who reacted badly to being excluded showed increased activity in two regions of the brain associated with physical pain, the dorsal anterior cingulate cortex and the anterior insula. Those who weren't bothered showed little or no increase in activity in these pain centres. The researchers suggest that because being connected is so important to the human species, that attachment system may have piggybacked onto the physical pain system over the course of our evolutionary history, 'borrowing' the pain signal to highlight when you are socially disconnected. This is why people talk about rejection as literally hurting.

Another study from a collaboration of the Universities of Michigan, Columbia, Colorado and the New York State Psychiatric Institute demonstrated that rejection activates brain areas that support the sensory components of physical pain – the secondary somatosensory cortex and dorsal posterior insula.[12] People who recently experienced an unwanted romantic relationship breakup were placed in the fMRI scanners while looking at photographs of their ex-partners and thinking about rejection. The researchers went on to compare these activated brain locations with a database of over 500 published studies. They found activation in these regions was 'highly diagnostic' of physical pain, with positive predictive values up to 88%. This study demonstrates that rejection and physical pain are similar not only in that they are both distressing but they also share a common sensory representation.

Physical pain is an alert system

When we lived in small nomadic groups, being ostracised from our tribes was a death sentence, as no one survived long alone. Our brains developed an early warning system to alert us when we were at risk of being pushed away by our tribe. Those who experienced rejection as more painful were also more likely to correct their behaviour to avoid getting kicked out, and therefore more likely to survive and pass along their genes. The pain heightened their stress arousal, stirring them into action.

Rejection is unique in this way. The emotional distress caused by other negative emotions such as shame or guilt is nowhere near as intense as the hideous emotional pain associated with most rejections. This is good news for a client to understand – the pain you feel does not mean you are weak, rather it is a natural response developed over millions of years. David was an intelligent, highly self-aware man, yet he still had doubts that his feelings were 'normal' in his situation. He felt rejected by the company for returning him to an unfamiliar environment, abandoned by his 'cronies' who had moved to new roles abroad and isolated from the people remaining around him.

Loneliness and depression

John Cacioppo, Director of the University of Chicago's Center for Cognitive and Social Neuroscience, told Fortune Magazine in a June 2016 interview that loneliness is not designed to be chronic.[13] Like physical pain or hunger, it's an aversive cue that alerts you to pay attention. It can also lead to depression which can reduce your desire to try to break back into the group. Depression sends a passive signal to the group that anyone who cares about you should come to your aid and reconnect. He believes depression can be adaptive in that sense. But social connection isn't just one way in receiving aid, it's about a two-way reciprocal, balanced relationship.

It's as likely that depression leads to loneliness as the other way around. Low self-worth is one of the biggest contributors to loneliness and a symptom of depression. If you place low value in yourself, it becomes challenging for you to make and maintain relationships. You don't see yourself as worthy, so there seems no point in trying. You may see others in an incorrect negative light, assuming that they must see as little value in you as you do yourself. So, you commence a self-perpetuating spiral – low self-esteem leads to isolation, and further isolation leads to lower self-esteem. This can lead to great unhappiness, alienating other people, putting them off, depression and at its worst, suicidal thoughts.

A Japanese study from Shimane and Hiroshima Universities showed that people with low self-esteem reported increased social pain relative to those with higher trait self-esteem, and scans showed a greater degree of dorsal anterior cingulate cortex activation.[14] In a computer game participants were exposed to social inclusion and social exclusion/ostracism. Researchers found significant connections between the dorsal anterior cingulate cortex and a network that includes the prefrontal cortex in the low trait self-esteem group. Low trait self-esteem appears to be related in the experience of social pain, and a connection between the dorsal anterior cingulate cortex and the prefrontal cortex may underlie this relationship. They conclude that low self-esteem predicts responsiveness to ostracism. Put simply, if you are suffering from low self-esteem, it is likely you will feel social rejection more painfully.

Loneliness and dementia diseases

Researchers at the Brigham and Women's Hospital and Harvard Medical School in the United States used findings from a study of 79 adults to examine whether 'cortical amyloid' levels in the brain – a marker for the early signs of Alzheimer's – was associated with loneliness.[15] They discovered that those with higher levels were more likely to be suffering from a higher degree of loneliness.

Older people who report feeling lonely are more likely to develop Alzheimer's disease and other forms of dementia, according to a Dutch study using memory and thinking tests.[16] The subjective feeling of being isolated and alone appeared to be a risk factor for Alzheimer's, regardless of whether someone was married or had a social network.

The researchers concluded that 'feeling lonely,' as opposed to 'being alone,' could be considered a major risk factor for Alzheimer's disease. It is not the objective situation, but rather the perceived absence of social connections, that increases the risk of cognitive decline. But scientists are unclear how or why feeling lonely may contribute to dementia risk. It is possible that feelings of loneliness may cause changes in the nervous system that dampen connections between brain cells, making the brain less able to protect itself against the onslaught of Alzheimer's. Alternatively, the fear and feelings of depression that may accompany the early onset of dementia symptoms may make people feel lonelier and more isolated.

While the study found evidence of a correlation between loneliness and early signs of the disease, it was unclear whether social isolation might actually be one of the reasons why Alzheimer's develops or if it was simply one of the symptoms. With an aging population, dementia represents a real health issue facing today's society and lonely feelings appear to be a contributor, an area worthy of further research.

Findings from solitary confinement

I live a few miles from Twickenham Stadium, the home of England Rugby and the largest dedicated rugby union stadium in the world. I recently saw the Rolling Stones there alongside 82,000 other fans – a huge crowd, and that number is around the amount of people in solitary confinement today (give or take a few thousand), across state and federal prisons, secure and restricted housing units, special management units and other isolation cells.

Although the practice has been largely discontinued in most countries, it's become increasingly routine over the past few decades within the American prison system. Once employed largely as a short-term punishment, it's now regularly used as a way of disciplining prisoners indefinitely, isolating them during ongoing investigations, breaking them into co-operating with interrogations and also separating them from perceived threats within the prison population at their request.

Psychologists and neuroscientists have attempted to understand the ways in which a complete lack of human contact changes people over the long term. At an American Association for the Advancement of Science's annual meeting, a panel discussion revealed that solitary confinement is irreversibly harmful to the mental health of the prisoners and not suitable as a means of rehabilitating them for re-entry back into society.[17] Craig Haney, a psychologist at University of California, Santa Cruz who's spent the last few decades studying the mental effects of the prison system, estimates that a third of isolated prisoners are actually mentally ill.[18]

Prisoners live in their tiny cells for at least twenty-three hours a day devoid of stimuli, denied physical contact and may go years without touching another person apart from a prison guard when restrained to go for exercise. Haney and colleagues' work through interviews and surveys shows, not surprisingly, that most prisoners suffer from severe psychological stress, dizziness, palpitations, and chronic depression;[19] 41% experience hallucinations and 27% suicidal thoughts.

Over time they withdraw from the tiny amount of social contact they have because any social stimulation makes them anxious.

Direct neuroscientific evidence is limited as prisons don't want their otherwise isolated prisoners to take part in research, however it is clear that as well as the effects referred to earlier in this chapter, the lack of exposure to the sun plays a role. Some brain activity is driven by circadian rhythms, which are set by exposure to daylight. Research has shown that restricting exposure to sunlight, and therefore interfering with circadian rhythms, increases the possibility of depression. There is evidence that circadian disruption alters the function of brain regions involved in emotion and mood regulation.[20] Brain architecture can also change over time. The hippocampus, which is involved with memory, decision-making and spatial navigation, has been found to dramatically shrink in the brains of people who are depressed or stressed for a long time.

Isolation at work

Research from Sacramento University and Wharton School of Business found that estrangement, alienation and lack of security in the workplace led to attentional deficits, poorer team contribution and withdrawal from relationships in the work place, leading to lowered performance.[21] When employees experience work loneliness, they are likely to avoid the possible stigmas associated with feeling lonely at work and take on a more defensive approach in their interactions with their colleagues. This could lead to surface acting, a type of emotional regulation where people hide or mask what they feel and modify their emotional expressions in a way that is inconsistent with their actual feelings. They found that loneliness triggered emotional withdrawal from their organisation which was reflected in this increased surface acting.

While loneliness may be thought of as a private emotion, the researchers highlighted that employee work loneliness is also a social phenomenon, as it is seen by colleagues and impacts team contribution and team effectiveness. This could lead to an increasing negative emotional spiral of loneliness as colleagues are likely to withdraw further as a result of their perception that the lonelier employees are less emotionally committed, therefore offering them even less connection.

It was useful to discuss this research with David. He had always prided himself as an inclusive leader of high-performance teams and believed he needed to be a role model. While he felt he was trying to be in his words 'his best self' at work during this lonely time, the concept of surface acting flew in the face of his value of authenticity. He decided to organise a team event in the near future where he and his new team could get to know each other in an informal setting and to start to build trust, collaboration and community.

The researchers suggested that management shouldn't treat work loneliness as a private problem that employees should deal with themselves, but rather should consider it as a larger issue that needs to be addressed both for the employees' sake and for the organisation as a whole. The social context of work is important and it can significantly shape employees' experience and behaviours. Work on the employee work experience, job design, cross department collaboration and team building are all areas that can help.

In/out-groups

Social categorisation is one explanation for prejudicial attitudes and a mentality of 'them' and 'us'. You see the group which you belong to, the in-group, as being different from the others, the out-group, and members of the same group as being more similar to each other. These manifest as harmless as different teams at a school quiz night 'booing' in fun when beaten, to more harmful behaviour like fighting breaking out at a major premiership football match. My son, who studies at Manchester University, quickly learned that his Spurs scarf should have been left at home in London. History offers the worst examples of in-/out-groups through genocides in Germany, Poland, Rwanda and Yugoslavia to name a just a few where the out-group suffers the most appalling atrocities.

David Eagleman, author of *The Brain: The story of you*, says that an understanding of in-groups and out-groups is critical to understanding our history of genocide.[22] Interviewing a United Nations translator in

Sarajevo, he learned that people his family had been living with for decades were capable of killing their own school friends. Through his neuroscientist eyes, he explored how it is possible for this to happen when certain situations short circuit the normal social functioning of the brain. In the lab people's brains showed a larger empathetic response when they saw a member of their in-group in pain (their hand being stabbed with a syringe needle) than when it was a member of their out-group. Given that there were just single word labels (Christian, Jewish, Atheist, Muslim, Hindu or Scientologist) he showed it takes little to establish group membership. A basic categorisation is enough to change your brain's pre-conscious response to another person in pain. This was not about the divisiveness of religion – even atheists showed a larger response to pain in the hand labelled atheist and less of a response to other labels. Eagleman says it's not about religion, it's about what team you are on.

In a study from Liverpool University, UK and Georgia Gwinnett College in the United States, fMRI scans were used to examine sensitivity to social exclusion when looking at faces of different races.[23] They found that participants demonstrated greatest dorsal anterior cingulate cortex activation when being excluded by self-resembling and same-race faces, relative to other-race faces. The anterior cingulate cortex is known to be a key player with physical and social pain. Additionally, participants expressed greater distress and showed increased anterior cingulate cortex activation as a result of exclusion in the same-race condition relative to the other-race condition. A positive correlation between implicit racial bias and activation in the amygdala was also evident. The researchers concluded that implicit attitude about other-race faces partly explains levels of concern about exclusion by out-group individuals. These findings suggest that for people who are more distressed their brain acts as a neural alarm system and shows greater activation when being excluded by individuals who they are more likely to share group membership with.

Some neurons can sense loneliness

A Massachusetts Institute of Technology and Imperial College London group used mice with a label on their dopaminergic neurons, the ones involved in reward behaviour, to investigate loneliness.[24] Dopaminergic neurons in the dorsal raphe nucleus of lonely mice exhibited an increase in synaptic strength compared to social mice – their synapses became stronger in response to loneliness.

The researchers performed a follow-up experiment to examine the neural basis of social rebound – the desire to join the social circle after a short time in isolation. A mouse that had been isolated was allowed to socialise with another mouse, which significantly increased the activity of the dorsal raphe nucleus dopaminergic neurons compared to group-housed social mice. These results suggest that as well as sensing loneliness during isolation these neurons also fire away when the mouse bounces back into a social rebound. In a third experiment, when stimulating the dorsal raphe nucleus neurons, the mice became more social; then when inhibiting them, the mice didn't spring into a social rebound after a lonely period. The conclusion of the study was that the dopaminergic neurons of the dorsal raphe nucleus are involved in motivating social behaviour. Similar conclusions after looking at serotonergic neurons instead of dopaminergic neurons in the dorsal raphe nucleus in response to social isolation have been found by researchers at the University of Toronto.[25]

Clearly, loneliness in humans is a subjective experience and is not as easily explored as with mice. This study highlights the importance of the dorsal raphe nucleus brain area as a new field of study for the neurobiology of loneliness.

The age of social media, likes and FOMO

Every second you access and contribute to social media, you open yourself up to being rejected, criticised, snubbed or pushed away. Being unfollowed on Twitter, teased on Facebook, friends not liking your latest Instagram post or not matching on a dating site are becoming more frequent occurrences that can result in a roller coaster of emotions. And the ever-present FOMO (fear of missing out) and wanting to be part of the in-group means you find it a challenge to switch off. Social media and constant contact to millions of people at any moment inevitably means that you open yourself up to more rejection, even if it's as small as someone not liking your social media post when you liked theirs. In addition, your online presence may by default create less direct, voice-to-voice or face-to-face contact with your personal connections.

University of Michigan research suggests that not only does the brain process rejection like it does physical injury, but that personality traits such as resilience are vital to how we process the pain of rejections.[26] In an online experiment similar to online dating, study participants chose profiles of people they were most attracted to. A PET scanner tracked their brain's opioid response, measured by looking at the availability of μ opioid receptors on brain cells, when they were told the person they'd chosen did or did not like them. The brain's natural painkilling response varies between humans, with some releasing more opioids during social rejection than others, meaning that some have a stronger resilience – or more adaptive protective ability. When μ opioid is released, there is a trigger in two areas of the brain. The amygdala processes the strength of the emotion, and the pregenual cingulate cortex determines how your mood changes because of the event. Therefore, the more opioid released, the greater reduction in pain and possibly a greater experience of pleasure when someone feels that they've been socially accepted or validated.

The participants had been told that the 'dating' profiles were not real, and neither was the 'rejection.' However, the simulated social rejection was enough to cause both an emotional and opioid response. A step further, and it could be argued that those prone to social anxiety, panic attacks and depression release less opioid, and therefore take longer and do not recover as well from negative social experiences. These individuals may also struggle to gain as much pleasure from social support as those who get more opioid in the pregenual cingulate cortex.

Social media is a double-edged sword. For David, it was a useful means to reconnect with old friends and to join with groups who shared his interests. But it doesn't replace human contact – and opens you up to a whole series of potential rejections and criticisms which you may not have experienced if you hadn't clicked the link.

Dealing with loneliness and rejection

As with many emotions and areas we come across with clients in coaching, a first stage is often just awareness that something is going on. For David, simply using the word lonely, with all the connotations it had for him, showed great strength and willingness to do something. Recognition that the genuine pain of loneliness is a sign that something needs to change to alleviate that pain can be the beginning of a client's journey. Through our coaching conversations David began to understand the effects that loneliness was having on his life both physically and mentally, telling me that it didn't make the mountain less slippery but it made it less steep.

Gratitude

As briefly mentioned in the previous chapter, I have my own gratitude journal, encourage my family to record experiences they are grateful for and discuss gratitude frequently with my coaching clients. I find writing down what I am grateful for to be a useful, sometimes humbling and cathartic experience which has helped me in many situations, including the dark days after bereavements and loss. Many self-help groups and organisations such as Alcoholics Anonymous regularly practice grateful thinking believing it to lead to enhanced psychological and social functioning.

Gratitude is an area of increasing interest to neuroscientists, not least because of its positive effects on well-being and health. I have highlighted some varied research in the following paragraphs.

Using a web-based survey with a convenience sample of 197 participants, a University of Rome study showed that gratitude seemed to moderate feelings of loneliness and highlighted the potential benefits of gratitude for strengthening social bonds.[27]

Two longitudinal studies across the Universities of Warwick, Nottingham, Coventry and Leicester, found that gratitude led to higher levels of perceived social support, and lower levels of stress and depression.[28]

Researchers from the Universities of California and Miami randomly assigned young adults to keep a journal of things they were grateful for or things that hassled them, or social comparison – reasons why they were better off than others.[29] They then kept daily or weekly journals. Participants assigned to keep gratitude journals showed greater increases in optimism, enthusiasm and energy and more time exercising compared to the other groups. Daily journaling activities were more powerful in facilitating gratitude. The researchers highlighted that downward social comparison was not as effective as other routes – being aware that other people are worse off than you is not gratitude. Gratitude requires an appreciation of the positive aspects of your personal situation. Sometimes noticing what other people don't have may help you see what you can be grateful for as a start point, but you have to take that next step. You actually have to show appreciation for what you have, for it to have an effect.

A University of North Carolina study shows that the reciprocal practice of gratitude has been shown to increase feelings of social inclusion and closeness in interpersonal relationships.[30] The neurotransmitter oxytocin is thought to play a major role in promoting close social bonds through social interactions. Oxytocin is released during hugging, touching, and orgasm in both genders. The study looked at whether couples in a romantic relationship expressing gratitude to each other bound them closer together. Taking a genetic approach, they showed that expressed gratitude could be associated with variation in a specific gene (CD38), known to affect oxytocin secretion. Their consistent pattern of findings suggests that the oxytocin system is associated with solidifying the 'glue' that binds adults into meaningful relationships.

David became a master at using his journal for recording his lonely thoughts, any self-sabotaging behaviours and gratitude for what he had in his life and work. At the beginning of his coaching, he told me he wanted two journals. One for planning ahead and creating a more positive future and the other for looking at his past. I was happy to indulge him and throughout our months of work together I noticed he visibly changed his body language when he grabbed his red (future) journal to make notes. He became more animated, purposeful and inquiring. When he accessed his grey journal (past) he was more studious and serious. He told me he was grateful for both of his journals as their distinct different purposes reminded him how he was proactively making changes to his life.

Giving back and volunteering

As humans are social animals, it isn't surprising that we are wired to help one another. Neuroscience demonstrates that both the giver and the receiver benefit from the relationship and that giving is a powerful pathway for creating more joy and encouraging overall health.

A University of Georgia review of neuroscientific research shows that altruism is associated with a specific pattern of brain activity.[31] The tendency to engage in altruistic behaviours is associated with greater activity within limbic regions such as the nucleus accumbens and anterior cingulate cortex, in addition to cortical regions such as the medial prefrontal cortex and temporo parietal junction. Much of this taps into instinct – we know when we see a child in trouble, our instincts kick in and we spring into action before we can even think.

A recent study from the Universities of Pittsburgh and California suggests that giving social support to others may neurologically benefit the giver more than the receiver.[32] Participants completed three tasks in the fMRI scanner (a stress task, an affiliative task and a prosocial task) and were asked about various scenarios where they either gave or received social support. For example, having someone to lean on or looking for ways to cheer people up when they were feeling down. The neuroimaging scans showed that giving had greater brain benefits than receiving. Specific findings were:

- Reduced stress-related activity in the dorsal anterior cingulate cortex, right anterior insula, and right amygdala.

- Greater reward-related activity in the left and right ventral striatum.

- Greater caregiving-related activity in the septal area.

The researchers concluded that these results help offer evidence that social support giving can benefit health.

To advance understanding of the amygdala, a team from the Universities of Yale, Duke and Pennsylvania studied the social behaviour of rhesus macaques.[33] They observed how these primates made beneficial prosocial decisions when playing a modified dictator game, in which one individual can donate or withhold rewards from another. The researchers discovered that neural activity in the amygdala directly mirrored the value placed upon generosity, kindness, and charitable behaviour. Neurons signal the value of rewards for self and others when monkeys make social decisions. These value-mirroring neurons reflected a monkey's tendency to make prosocial decisions on a moment by moment as well as long-term basis. The scientists could actually predict when certain monkeys were going to be generous and charitable based on their neural responses.

They also discovered that when oxytocin was introduced into a specific region of the amygdala, prosocial behaviours and attention to the recipients increased instantaneously.

Of course, more research is needed; however, like humans, the stronger the bonds these monkeys have, the more successful they are. Ultimately, monkeys with more close friends live longer and have more offspring.

Giving back, volunteering, caregiving and altruism can be helpful for lonely people. The positive energy you feel from contributing releases endorphins and yields a feeling of satisfaction. It also catalyses gratitude, helping you to focus on what you do have rather than don't have, helping to ground you, acting as a distractor which in turn can dilute your emotions about your own situation.

Clearly there isn't a direct causal relationship. Giving back is not going to cure loneliness in one shot, of course not. And not everyone benefits. You may be feeling stressed and think adding some volunteering into your already time-strapped day will just make it another task on the list. Which of course defeats the purpose. Sometimes in sad or challenging caregiving situations you can find yourself over involved and buried in sad emotions which has a negative effect. Burnout is common among people in caring professions and they can end up suffering from what is called compassion fatigue. It is imperative to always look after your own physical and mental health before taking care of others. Abrupt though it sounds, I recall vividly saying to my Father a few years ago, 'Dad, you are no good to Mum if you are in a coffin.' I said it to shock him into some action as, out of love and devotion, in his early eighties, he believed he was the only person who could look after my mother, who was suffering from the later stages of Parkinson's. It was clear that he needed to remain as healthy as he could, both physically and mentally, so he could give Mum his best self when he was with her.

Do you have a best friend at work?

Just before I started coaching David he completed the annual employee engagement survey. He was struck by his reaction to the question 'Do you have a best friend at work?' and realised his loneliness at work was taking its toll on his attitude, motivation and health. In our work around Employee Engagement at MyBrain International, we often cite the Gallup 12 questions, the most controversial of which is the one that gave David pause.[34] Gallup's employee engagement work is based on more than thirty years of in-depth behavioural economic research involving more than 17 million employees, across multiple levels in companies, sectors and geographies. Through their rigorous research, Gallup has identified twelve core elements that link powerfully to key business outcomes and best predict employee and workgroup performance.

In the most enjoyable places to work, employers realise that people want to have quality relationships with their colleagues, and that company loyalty can be built from trust. Gallup recognise that the development of trusting relationships is a significant emotional connection for employees in today's marketplace and is key to retention. They say that the best managers understand that the quality and depth of employees' relationships is a critical component of employee loyalty.

The element of 'I have a best friend at work' is challenging to answer for many people who get stuck on the word 'best,' because they feel the term implies exclusivity. Gallup trialled different language but found without the word, the item lost its power to differentiate between highly productive and mediocre work groups. Their research showed that the item of 'best friend' matters for overall engagement. After defining his presenting issue of loneliness as triggered by the company engagement survey, I later introduced the Gallup research to David. We used it as a platform to talk about trust and support as well as the wider area of role, team and company engagement. It added a useful dimension to deepen our conversation and provide an application level to the neuroscience David had become so fascinated with.

Gallup has found that compared to those who don't, employees who have best friends at work identify significantly higher levels of healthy stress management, even though they may experience the same levels of stress. This made sense to David, as he felt he had lost his resilience and coping 'muscles' since his trusted colleagues had moved on.

Recognise you have strengths – don't assume people think badly of you

One of the dangers of loneliness is the spiral of feeling not worthy. You feel excluded and not part of the team and tell yourself it must be something you have done wrong. Every time you think about a past perceived rejection your self-esteem takes a knock and you become self-critical almost looking for areas where you can blame yourself. There's nothing like kicking yourself when you're down, boy it feels bad.

When giving feedback on our brain-based MiND tool, we always help clients to gather a language for their preferences and strengths (see Appendix for details of the MiND tool). We come from the perspective that development is as much about learning new skills or adding a layer to existing skills as it is about using strengths in different contexts. We took a couple of our coaching sessions to explore David's strengths of spontaneity and emotional intelligence culminating in him writing a letter to himself expressing gratitude for each. Importantly he was able to define why each strength mattered to him, how it contributed to his self-image, examples of where he had used each resulting in a positive outcome or success and how he used each in his work and personal life. In so doing he experienced a form of self-validation that he had a lot to offer, not just for himself but for others as well.

Encouraging feelings of social connection

Our social groups are often a result of our circum-stances. I am still great friends with a few of my old flat mates from my first hall of residence at university where we were randomly allocated rooms. I see some Mums from my son's junior schools regularly for coffee even though the boys are all at university or working.

I joined a local running club to sort out my fitness and have become friendly recently with a great group of people – all different ages, capabilities and with different running goals, but bound by our desire to improve and make our running experience fun. But life moves on, I am no longer connected with all of my old flat mates from the 1980s or all my son's friends' Mums. The larger group worked at the time because of the circumstances and context.

Just looking at old photographs of people you care about can provide a social 'snack'. Social snacking is a strategy proposed by Northwestern University researchers that is useful when direct social connections are unavailable.[35] It refers to indirect strategies like using tangible symbols such as a photograph or gift someone you cared about gave you that provides the illusion of connection. As food snacking can ease hunger for a short time when you can't get a proper meal, snacking on reminders of important social connections can ease your social 'hunger' when you feel alone. In these days of hot-desking and home working, David was unusual in having his own office. It was a lovely space with windows and a view but impersonal, bare and frankly soulless. He shared with me that prior to his ex-wife's death he had family photos on his desk and his filing cabinet, as well as golf trophies and pictures of work award accolades. On returning to work, he hadn't felt he wanted to display them again and kept them in two drawers in his desk. I encouraged him to talk about them, what their significance was

LONELINESS HURTS

and why he had originally selected each to take to his work space.

Connect with people with similar interests

We discussed the need to focus on developing and sustaining quality relationships with people who share similar attitudes, interests, and values to ensure he kept his loneliness in check. David was convinced his loneliness was a work-based issue and not a home-based one. He had received a lot of personal support since his wife's death and spent every Saturday with his 'cronies' at his golf club. The photograph conversation however revealed a past passion for the sport with pictures of David with customers, David with colleagues, David with suppliers, David with his team – all on what looked like top-level golf courses. In past days he had been captain of the company golf society which disappeared when he took his two-year assignment in Singapore. On return, events had taken over and golf didn't appear anywhere on his agenda for work, 'I hadn't given it a thought.' It's interesting when a solution is right under your nose but you delete it because your attention is elsewhere. Initially closing down the idea that came up as it 'will be like pushing water uphill in the current climate' David decided to gather some views from people he knew to gauge interest and then to approach HR to see what support the company could offer in terms of funding and time to restart the golf society.

Go online

Social media gets a bit of a caning in the press and in self-help books. As mentioned earlier in this chapter, it is easy to see why your participation in social media can fuel loneliness or feelings of rejection. Everyone you know seems to be having a happier life than you, looks more beautiful, has more connections, is on a beach with bluer sea or having fun at an event you haven't been invited to. But the underpinning ethos of social applications is about connecting people. If used with that intention, portals like LinkedIn and Facebook can open up your world.

When I left corporate life fifteen years ago, I knew a potential derailer for me would be lack of social contact, lack of access to a team and a sense of community. At the time social media was in its infancy so I joined other entrepreneurs, coaches and trainers by becoming an associate of larger training organisations. This enabled me not just to keep learning, but to be part of something larger, to network and achieve a sense of belonging. Since those days, face-to-face networking groups have sprung up all over the place and the proliferation of social media communities make the transition to self-employment much easier. As well as working with a business partner who shares my passion for neuroscience, who I learn from every day and I trust implicitly, I am an active member of seven value-adding Facebook Groups and a similar number on LinkedIn. I feel connected to a community of like-minded professionals, can ask questions, raise points and have a voice.

Through joining the 'modern world' in his words, David was able to reconnect with a number of people, from school and university friends to old work colleagues and even distant relatives through his new hobby of building his family tree. I encouraged him to expect the best and be excited about possibilities that could emerge. Lonely people often expect rejection, so instead David focused on positive thoughts and attitudes in his social relationships. It took initial mental effort and some resilience along the way but for David, his social media journey has reaped dividends and acted as an antidote to his symptoms of loneliness. He even discovered a man who he had shared a flat with in his university hall of residence who worked for the same company four floors above him. Oh, how anonymous the corporate world can be.

Identify self-defeating negative behaviours

Self-sabotaging behaviours are challenging to see in the moment but easy to see the next day. When our son first started school and we entered the ritual of the whole class being invited to birthday parties, our weekends were filled with football, ceramic making, bouncy castles and pirate parties. At the beginning I stayed with him but a few years in (until he was eight) he still asked me or his Dad to remain with him at the party. While loneliness wasn't driving his behaviour, rather he was a shy little lad, the point is that he was ready to join in just as the party was ending – or not at all. The next day he would always say to me 'I wish I'd played football with Christian', 'I wish I had sat next to Nathan', 'I wish I had worn my pirate outfit'. Hindsight is a wonderful place to recognise what sabotages you.

You may feel justified standing on your own at the obligatory buffet at the boss's house if you see your colleagues enjoying themselves in conversations with others. But the next day your social anxiety will have diminished, allowing you to understand that was what was driving your behaviour. Social anxiety and telling yourself you were right, no one was talking to you.

Loneliness can make you generate negative thoughts by thinking about social events. You get invited to the office Christmas party and immediately think about being on your own, not knowing what to talk about, who to hang out with, thinking the other girls will laugh at your lack of fashion sense. The thought of being stuck with boring Maurice from Marketing fills you with dread and panic and all you can think of is impending disaster. I suggested to David that he visualised successful realistic results from his inter-actions by reminding himself of people he had had positive communications with in the past. Positive interaction does not mean that they asked to be your best friend. It can be as simple as a cheery hello as they opened a door for you. Thinking realistically is about overcoming the beliefs that your loneliness is bringing, where you imagine no one will talk to you or want in any way to associate with you and you will be outcast from every group in the room. Unless you have a reputation for being a psychopathic murderer, in reality, that just won't happen.

Learn to catch your negative chat. If you have taken some action to engage socially with others and you haven't heard back from someone, it does not necessarily mean they don't want to spend time with you. It could be that they didn't receive your email, they have gone to a Zumba class after work and haven't picked up your message, or they are hunkering down in bed with the flu and a box set. Confirmation bias reigns in these situations. Five minutes of someone not responding and you tell yourself 'told you so, knew that wouldn't work'. Give people the benefit of the doubt and remember, building relationships takes time.

When you feel lonely it can be tough to act and push yourself to connect to others, and it is understandable that your initial courage can quickly give way to fear of rejection. But tell yourself, indulging these negative feelings will only set you spinning back into your vicious circle and bring about the very thing you are working to avoid.

Four-legged friends

David owned a labrador, Alfie, which he referred to as the love of his life. He told me one of the reasons he wanted his ex-wife to live out her days at his home was so she could benefit from Alfie's company. He had taken a six-month leave of absence so on return to work he missed his four-legged friend, making his perceived isolation more acute and he organised his life to work from home as often as he could so he could have company he enjoyed. It was a good strategy in one sense as it brought him comfort, however working at home didn't help to deal with the real issue of his loneliness at work.

Pet therapy is a field of increasing interest, particularly in the care sector for the elderly. A recent joint study by the Universities of Manchester and Southampton concluded that pets should be considered a main rather than a marginal source of support in the management of long-term mental health problems, which has implications for the planning and delivery of mental health services.[36]

David now

In catching up with David to seek his permission to include his story in this book, he told me his work situation had improved, having run regular team sessions, recruited some like-minded colleagues into his team and becoming part of a mentor team for young high-potential employees. He felt more involved, his opinion was proactively sought and he had positive relationships both within and outside his department. The golf society was going so well that in his new European role, he was organising a cross country league. He shared that Alfie had a labrador girlfriend Meg, whose owner was potentially open to romance too. Work was no longer a lonely place and he felt fulfilled and hopeful in all walks of his life.

CHAPTER FIVE

Decisions, Decisions, Decisions

'In any moment of decision, the best thing you can do is the right thing, the next best thing is the wrong thing and the worst thing you can do is nothing.'
  – Theodore Roosevelt, 26th President of the United States

We make decisions all the time every day, some through habit that we are barely aware of and some which require conscious weighing up of choices and options. Some decisions are easy, some make us anxious, some we just make to get on with our busy days and some we ignore or delay. Whether in the supermarket noticing discounts, in a restaurant, the Starbucks queue or flicking through Netflix, we identify potential options, compare their values and make a choice. We make decisions both consciously and subconsciously using a variety of strategies such as balancing alternatives, acting on impulse, agreeing with someone else's decision, delegating the decision, avoiding it, voting or working to consensus in a group, or taking time to consider by sleeping on it.

I can't find a source for the amount of waking decisions made every day but according to Cornell University researchers, we make 219 decisions each day on just food and drinks so total decisions in a day must be huge.[1]

Having a better understanding of the neuroscience behind decision-making may help you make better decisions that lead to positive outcomes for your future. You have the ability to control your decisions and possess the power to change your mind and make a different choice at any time. Most of the research on decision-making focuses on attempting to understand why people in the same situation make different choices and what happens in their brains leading up to the final choice.

As mentioned before, neuroscience has become popular in recent years since brain scanning technology has enabled more knowledge – and indeed since communications has opened up the world in sharing and discussing these new insights. In the many 'neuros' that have developed, neuroeconomics is an area which has greatly contributed to understanding decision-making and how it manifests differently for different people and situations. Much of the research looks at decision-making in ambiguous, uncertain environments and is explored in a laboratory setting with experiments simulating this ambiguity. Research provides measures of decision-making behaviour, often assessing risk-taking tendencies, the possibility of reward or exposure to negative outcomes. Many experiments use gambling or lottery tasks.

LAURA'S STORY

Laura worked in the automotive industry as an events manager with a team of ten. She was responsible for product launches, events for customers and the press and for organising trade shows across Europe. It was a highly visible role and a busy job she thrived on. She enjoyed the pressure, the many moving parts, the profile and exposure across the business. Even though the role required regular presentations and meeting new people, Laura considered herself an introvert and prided herself on her ability to think things through and reflect on her decisions, drawing on her experience and that of her team. She was generous with her team, enabling them to learn on the job through regular coaching and always giving them credit for a job well done.

She came to me a few months after her reporting line changed after a reshuffle of the senior leadership team. She had applied for her manager's role and had been given the feedback that she was a strong contender but deemed not ready. She was determined to learn from her new manager Paul and although she liked him and his open, collaborative approach to the team she felt he was piling on the pressure. In her regular one-to-one meeting Paul told her that as well as her, he had identified two other colleagues who he felt were high potential and he was giving it six months before selecting one to be his back-up and formally named on the company succession plan.

Even though she was delighted she had been recognised, Laura hadn't been prepared for a 'competition' and she was cross that she had allowed herself to assume the role of number two. At first she kept a positive attitude, believing that her performance and results would win the day but at a team meeting she realised that both of the other contenders – Gail and Richard – were approaching the potential role of Paul's successor like a marketing campaign and were clearly going the extra mile in their work. They were good friends, 'rampant extroverts' and posturing for position. She felt Paul respected Gail particularly, listening to her input at meetings with enthusiasm and encouragement.

Gradually Laura had begun to feel marginalised and lose confidence in her ability and her place as one of Paul's key players. She recognised this was her own issue to sort out and that fundamentally Paul was a good and fair manager. Her biggest concern was her erosion of confidence in her decisions. She had become hesitant and had missed a couple of deadlines. One of her team had saved the day by covering for her at a meeting where she wasn't prepared. That was the final straw for Laura and she proactively sought out a coach, at her own expense. When she came to me, she was confused, felt she was going in circles with key decisions, stuck about potential next steps, and lacking in confidence about her ability to do her job.

The brainy bit

Decision-making is a complex process, and individual brains are uniquely different. Your past experiences and memories, your specific preferences, personality, the decision context, levels of perceived threat and reward, your risk-aversion, how you respond to ambiguity and incomplete information are all factors in your decision-making strategies.

In making decisions, you rely on working memory to help you weigh up your different options, and you might access long-term memory for clues from past experiences. Simply recalling a particular memory may trigger an emotion and fire up your limbic system. At the same time, the cortical attention areas of your brain are used to keep attention on the decision you're trying to make, adding in some rationale. It takes effort and concentration to use all the different parts of the brain involved in making sense of all the factors which need to be considered to get to the decision.

No single part of the brain can be highlighted as the 'decision-making centre'. The research discussed in this chapter includes studies that are designed to show specific parts of the brain involved in decision-making per se and also brain regions associated with specific types of decisions, for instance risk and ambiguity. There is also a body of work demonstrating that working in harmony, intuition and rational thinking provide a 'checks and balances' approach to lead to a measured, informed decision. Decision-making involves a distributed network of brain regions including frontal, parietal and limbic structures.

Prefrontal cortex, decisions and eureka moments

In 2014, researchers at the University of Zurich showed that the interactions of neurons between the dorsolateral and ventromedial prefrontal cortex not only play a central role when you need to decide between several options during goal-directed, self-controlled behaviour but are also active during general decision-making.[2]

It was previously believed that the prefrontal cortex was only activated when self-control was required during decision-making between conflicting preferences. Decisions that require self-control are extremely important, as they can directly affect your well-being and it now appears that flexible, general decision-making also makes use of the same areas.

Researchers from the universities of Pittsburg, Washington and Stirling used fMRI scans to track brain activity in the lead-up to a decision by gradually revealing a photograph.[3] Participants signalled the time of recognition with a button press. Before the decision was made about what the picture showed (before they recognised the picture) activity registered across the brain in several occipital regions. Naturally, activity increased as more of the picture was revealed, suggesting a role of lower level sensory processing. In the inferior temporal, frontal, and parietal regions, a gradual build-up in activity occurred, peaking at the time of recognition, suggesting that these regions participated in the accumulation of evidence. The medial frontal cortex, anterior insula, and thalamus remained at lower level activity until the moment of recognition, when activity spiked suggesting a relation to the moment of recognition or the decision itself. The researchers concluded their findings show a hierarchy of neural pathways at play in decision-making in which the 'eureka' moment is developed and builds through specific parts of the brain.

Decisions and subcortical areas

Hierarchical operations also seem to occur in the limbic system during decision-making according to a Japanese study, where researchers found that the striatum appears to divide in turn into three different sub regions.[4] They looked at distinctive functioning within the striatum in decision-making with rats performing fixed and free choice tasks by investigating information represented by phases of active neurons in the dorsal lateral striatum, the dorsal medial striatum and the ventral striatum. The rats were required to poke their nose in either a left or right hole after a cue. A food pellet was given depending on the cue and the action (the decision) the rat took. The probability of the reward was fixed in one task and varied in the free choice task.

The researchers found the three parts of the striatum worked together in a co-ordinated hierarchy, ultimately collaborating in different phases of decision-making. The ventral striatum was most active at the beginning of a decision-making process. During active selection, the dorsomedial striatum changed firing levels, when the rat was considering the expected reward or consequence by making the decision to poke left or right. Lastly, the dorsolateral striatum fired short bursts at varying times throughout the task, suggesting it is gearing up the motor movements required once a decision is made and action is taken.

The findings suggest that the rats in the experiment analysed the potential benefit of choosing the left or right poke during the dorsomedial striatum phase. To the researchers' surprise, there was little difference in the hierarchy of striatum firing during fixed or free choice. The researchers proposed the hypothesis of hierarchical reinforcement learning in the basal ganglia to explain the results which doubtless will open up more research avenues.

Risk

Most decisions come with some risk. It is useful to be aware of the underpinning neuroscience of whether you are more risk averse or risk seeking so you can consciously consider your risk bias in future decisions.

Researchers at the Universities of Bonn and Zurich showed that choices are made dependent on the perceived likelihood and also the willingness of the person to take a risk.[5] Participants were grouped into risk averters and seekers according to preferences they revealed in a separate lottery task. The study found that during the anticipation of high-risk gambles, risk averters show stronger responses in ventral striatum and anterior insula compared to risk seekers. In addition, signals in anterior insula, inferior frontal gyrus, and anterior cingulate indicated that risk averters do not dissociate properly between gambles that are more or less risky than expected. The researchers suggested this may result in people overestimating prospective risk and lead to risk avoidance behaviour.

Laura felt that our discussions about her attitude to risk were useful, helping her to recognise that she perceived Gail and Richard as threats for the slot on the succession plan. This had likely contributed to the erosion of her confidence over recent months resulting in hesitation and avoidance of making decisions. Understanding that she was exaggerating the 'threat' or risk level of Gail and Richard's roles enabled her to reframe the situation, recognise that there would always be other great candidates for jobs she was interested in and she needed to reboot to bring the best of herself to her current job to re-enter the playing field. This discussion also helped Laura to be clear that she did want to become Paul's successor and it was the right career path for her.

Memories and decisions

Most decisions will be based to some extent on trial and error and your experience. Like habit formation (see Chapter 1) the more you make one type of decision, the more you will be able to assess the outcome, remembering which ones worked out well and which didn't in the past. You have set neural pathways laid down for repeated decisions which will help you to predict future outcomes and increase the likelihood of a good result from the next decision. So, decision-making is tightly linked with learning and memory and can be seen as a connection between recall of the past and future actions. A recent University of Minnesota special paper reviewed eleven different studies identifying the computational and neurophysiological processes that underlie these parallel memory and decision-making abilities.[6] The researchers report that where decision-making processes fail in terms of their trade-offs depends on the computational availability of memory representations. A memory representation that provides quick generalisation but little specificity is going to produce decisions that are fast, but inflexible, while one that provides many details, but requires considerable processing to break down and rebuild those details into a memory will produce decisions that are slow, but flexible. The researchers conclude that the same underlying neural systems that are critical for memory are critical for decision-making.

Decisions and emotions

As your experience grows with a particular type of decision for a particular context, you get better at making those types of decisions both consciously and unconsciously. As well as accessing memories, sometimes emotions can emerge in your conscious mind as a gut feeling or intuition that you can't precisely put your finger on. If it is a particularly strong gut feeling, it would serve you well to pay attention to its message. The quality of your decision-making will be based on varying amounts of experience that have been previously tried, tested and assimilated. That expertise increases your chances of getting it right through being 'informed' by intuition from familiar neural pathways. Using your conscious prefrontal cortex to 'listen to' your intuitive feelings can lead to better decisions. If used in harmony then you are giving yourself the best chance of success.

Trust your gut instinct if you get a good feeling about a rookie local handyman with a great recommendation from a neighbour you like. Listen to your instinct about the job candidate who looks 'too good to be true' on paper and at interview you find out she has been lying. Pause with your emotional thoughts and call up your rational mind. You may be cross with a colleague and want to give her a piece of your mind but your rational mind knows that tomorrow, you may wish you had acted with more measure, no matter how compelling your desire was at the time. In the heat of the moment, emotions can lead you to make choices that hurt your long-term interests – shouting at your child, spending money on credit, flirting with a stranger in a bar on a work trip.

Of course one of the functions of emotions is to guide you towards pleasure and away from pain. With decisions it's critical that you're able to judge what risks are worth taking – and emotions can help you make those judgements. Working in harmony, your emotional limbic gut together with your rational prefrontal cortex provide the 'checks and balances' approach that can lead to a more considered, better decision.

Risk and the gut

In today's ever-changing world, there will always be uncertainty about outcomes in much of your life. Sometimes you can estimate the risk and factor that into your decision; however, if there is an ambiguity of risk, that can present more challenge. Researchers at the University of Kentucky examined the neural basis of decision-making under different types of uncertainty – ambiguity and of unknown outcomes.[7] They found different ways the brain processes decisions in these conditions. For those tolerant of ambiguity, the decision involved parasympathetic systems including the lateral anterior insula. This was associated with gut processing of the information. But for participants with ambiguity aversion, decisions with missing information were dependent on processing in the prefrontal cortex so activating logical conscious decision-making processes.

Emotions at work

Years ago, when I worked in the corporate world, I remember a manager in the business who was adamant that there was no place for emotions at work. He argued that work should be a place of logical, rational thought, where you don't 'give in to' emotional thinking. And you absolutely shouldn't display any emotions as it would be completely unprofessional. It certainly made for a black and white culture for his team. Daniel Goleman caused a sensation in some working circles with his book *Emotional Intelligence: Why It Can Matter More Than IQ* in the 1990s.[8] Despite many business leaders quickly embracing the concept, with some of my clients it still seems that rational thinking is aspired to, respected and maybe even rewarded at the cost of emotional intelligence. This can have a huge impact on relationships, respect and trust at work, creating a command and control culture more reminiscent of organisations forty years ago.

University of California, Los Angeles Professor of Neuroscience Antonio Damasio, explains that feelings arise as the brain interprets emotions, which are the physical symptoms of the body reacting to external stimuli. His long, respected career includes research around emotions and decision-making, memory, language and consciousness. In his seminal book *Descartes' Error*, he says that effective decision-making is not possible without the motivation and meaning provided by emotional input and introduces readers to his patient, 'Elliot'.[9] Previously a successful businessman, Elliot underwent neurosurgery for a brain tumour and lost a part of his brain – some tissue in the orbitofrontal cortex, that connects the frontal lobes with the emotional brain. He became a real-life Mr. Spock from Star Trek, able to function but with no emotions. But rather than this making him perfectly rational, he became paralysed by every decision in life. Elliot could think up lots of ideas, but he couldn't choose. Damasio commented that he started to think that the cold-bloodedness of Elliot's reasoning prevented him from assigning different values to different options, writing that his decision-making landscape became hopelessly flat.

Damasio later developed the somatic marker hypothesis to describe how visceral emotion supports your decisions. Originating in the insula and the amygdala, these markers send you messages that something feels right, or not. The more you pay attention to trusting your intuition in combination with facts, the better your future decision-making can become. He argues that the semantic markers act as a filter and a shortcut to decision-making, reducing alternatives and choices. While Elliot's world had 'flat' values, most people weigh up potential outcomes that are left after the somatic markers filter the other possibilities out.

Damasio tested this theory in his famous experiment called the Iowa Gambling Task where subjects could choose between decks of cards to win money.[10] Among the choices were two decks that consistently led to accrual of profits and two others with riskier cards. Although the winning cards were worth more than the winning cards from the safe decks, the losing cards were so damaging that, if chosen repeatedly, the risky decks would eventually bankrupt the player. Sweat levels were monitored and after drawing just ten risky cards, the body displayed signs of anxiety, meaning that their feelings were firing signals faster than rational thought. Even before they recognised at a conscious level that they had made a bad choice, the body was signalling information. Even though somatic markers aren't 100% reliable, it is useful to pay them attention and not ignore them.

In the game, he found that participants were initially attracted to the risky decks because of the large potential payoff. However, players soon veered towards the safer decks where they fared better in the long run. One of the test groups were neurological patients, like Elliot, with damage to the orbitofrontal cortex associated with emotional sensitivity to reward and punishment. Though their cognitive reasoning was unimpaired, they didn't experience the negative emotions that normally accompany large losses. Like the participants with no brain damage, these patients were initially attracted to the riskier decks, but because they failed to respond emotionally to large losses, they never learned to avoid the risky gambles. The lack of emotions didn't enable them to be 'informed' by any associated physical response.

Damasio, and colleagues from Stanford, Carnegie Mello and Iowa Universities, conducted a follow-up experiment with a variation.[11] This time, participants repeatedly chose between keeping or investing $1. If they invested $1, they had a 50% chance of winning $2.50 and a 50% chance of losing the invested dollar. In this game, it is best to always choose the risky option. Individuals who fail to invest, out of fear, suffer financially. As in the first experiment, players were initially attracted to betting on risky gains, but similarly, they became more cautious after experiencing some losses. In contrast the group of patients with orbitofrontal cortex damage continued to invest regardless of losses. In this task, the patients who didn't experience emotion outperformed individuals experiencing the fear of loss.

The lesson from Damasio's fascinating studies is that experiencing negative emotions can help and hinder decision-making. It all depends on the context and circumstances. Clearly you can't trust that your emotions will always steer you in the absolute right direction, but 'listening to' your gut and using what comes up to inform decisions you then consciously think about is a good strategy. Used together, emotions and rational thinking make good partners – you need emotions to make decisions, their input means you're not a cold, calculating Mr Spock operating in a Starship Enterprise life.

An interesting coaching session revealed that Laura trusted her gut instinct and that was part of the reason she liked to take time to reflect on decisions, so she could 'listen to' her emotions and take them on board. She thought the emotion of the discovery that other people were in the frame as Paul's successor, particularly so close to missing the top job at interview, had scrambled her emotions, getting in the way of her usual decision style. She had filled up her usual reflective space with, in her words, 'noise' from being so cross about the 'competition'.

Decision fatigue

Ever felt 'brain dead'? Sometimes life just seems like an endless list of decisions. When all three of our children were in school, weekends typically felt like a never-ending series of decisions, what homework should be prioritised, when to fit in music practice, would my husband or I drive to rugby practice, which child to help first for next week's test, what to have for meals, and slips of paper semi-guiltily being found from the school bag at the eleventh hour announcing book day or the requirement for a packed lunch. At the time I was grateful for school uniforms as it took away the need for any decisions about what to wear, which would have added to the 'decision debate' particularly with our two daughters. Even small decisions which aren't potentially life-changing like what to have for dinner, which coat to choose or whether to go to Sainsbury's or Tesco can wear you down and lead to what scientists call decision fatigue or ego depletion.

According to Bauminster and Tierney in their book *Willpower: Rediscovering our Greatest Strength* the adult brain may only represent 2% of your body mass but it consumes more than 20% of its energy for normal functioning.[12] And sometimes, like on my Sunday evenings, it felt like 100% of my brain was drained, with little capacity for anything other than retiring to bed.

Think of your brain like one of your devices. The more ways you use it, for apps, music, data roaming, voice calls, Facetime, texts, WhatsApp, Instagram, the quicker the juice will run out. One of my most successful purchases for Christmas for my teenagers (on a buy one get one free offer of course) was a special solar-powered gizmo that charged their mobile phones. Given that festivals were big on the agenda that year I became supermum for a nanosecond – my son and elder daughter reacted as though I had bought them a car each.

Using your brain is the same as using your smartphone at different intensities throughout the day. Some days you can comfortably work through until late in the evening, and then on other days, you feel 'brain dead' and shattered by lunchtime. It's about how you use your prefrontal cortex, how often you rest and restore it, how hard you have worked it, how much challenge you have expected of it and how many decisions you have asked it to consider.

If you've spent a lot of time focusing on a challenging piece of work, exercising self-control, or if it feels like my weekends did with school age kids, that you've made loads of choices, then you may not have the headspace to consider a life-changing or other major decision. Probably best to have a rest or change your environment if you are going to decide which house to buy, where to study at university or whether to propose to your girlfriend.

A University of Albany study showed from four experiments that even after people become depleted they can perform well if they are motivated to do so – but then they are much more depleted if self-control is quickly required for another task.[13] Like a tired marathon runner who finds some strength to sprint for the final 500m, they allocate their resources cautiously once they start to be depleted. The researchers also showed that people hold back more when they expect subsequent demands than when they think they are dealing with the final one.

In a series of experiments researchers from the University of Kentucky worked with leaders in powerful positions.[14] They found that power improves self-regulation and performance, even when the person is ego-depleted. When ego-depleted leaders work hard

they sometimes outperform non-depleted ones. However, the continued high exertion of leaders when
depleted takes a heavy toll, resulting in larger impairments later. Additional studies showed that leaders
sometimes look down on tasks they deem unworthy,
not putting in much effort, so performing poorly. This
is an important area for coaches to raise with clients in
leadership positions – their expenditure of self-control
resources and need for multiple decisions may help
them prioritise their efforts to ensure their goals are
met.

Decision timing

There is a famous piece of research, conducted in Israel
in 2011, where judges heading parole boards make
decisions about whether to grant parole, release prisoners and free them from tracking devices.[15] Most of
us view judges as fair and just, however of course they
are also human and experience the same daily rhythms,
sleep issues, work distractions, overload and stress as
everyone else. The researchers found that judges were
more likely to issue a favourable ruling in the morning
than in the afternoon, granting in favour of prisoners
65% of the time earlier in the day. As the morning
elapsed the rate of favourable rulings fell. A prisoner
was far more likely to be released on parole at 9am
than at 11.30am. Interestingly, immediately after the
first break for lunch they became more lenient and then
began to sink into the more tough decisions after a couple of hours. A mid-afternoon break similarly resulted

in more favourable rulings immediately afterwards. The researchers cannot be sure of a single specific causal reason for this pattern. Blood sugar levels could be replenished through snacks, mood could improve through moving and changing space – particularly if the judge took a walk outside – it could also be due to improving fatigue if they were able to rest for twenty to thirty minutes. Whatever the reason, whether and when the judge took a break was significant in deciding whether the prisoner was granted parole – or not.

Spending too much time deciding

Laura was concerned that she sometimes felt paralysed with decisions and took too much time. It was a mix of decision fatigue and also her desire to be sure she always made the right decision. As with all my clients, our coaching work included learning about strengths and preferences using the MyBrain International MiND tool (see Appendix for details of the MiND tool). Laura described our feedback session as eye opening, learning that her profile reported preferences across all quadrants. This meant that it was easy for Laura to step into many different worlds and see problems and solutions from multiple viewpoints. There is clearly an upside to this, however it may also result in vacillation with decisions due to the strength of seeing so many different perspectives. A disadvantage of seeing things from all points of view is that it can be hard to reach a conclusion. This was certainly the case with Laura, and the feedback discussion enabled her to reframe the situations where she found it challenging to decide. She recognised that this played out as decision avoidance in her manager's eyes and learning about her multiple preferences enabled her to put a strategy in place to help.

Research shows that consciously thinking through choices too much can actually impair decision-making. At the Max Planck Institute for Research on Collective Goods in Bonn, researchers across three studies showed that in some circumstances thinking too much can lead to a worse decision and that it might be best to let the 'unconscious' do what the researchers call 'the busy work' and trust your instincts.[16]

Decision fatigue explains why it is so easy to be distracted, go off at tangents, make mistakes, revert to bad habits and to avoid making decisions at the end of a hard day. You just have less stamina and have run out of battery life.

For Laura, she had invested a huge amount of energy in applying for her manager's role, then more energy handling the disappointment of not being successful. On top of that she then had to deal with her emotions of a changing landscape that included Gail and Richard. No wonder her head was full and it affected her decision-making. She then fell into a spiral of focusing on her 'competition', filling up all her cognitive reserve with concerns about the other two candidates which in turn impeded her ability to return to the high-performing leader she was in her role. Her battery was draining fast.

Cognitive biases

Decision-making is fraught with these biases that mask your judgement, many of which you are unaware of. I am thrilled to have recently become a great aunt to a beautiful baby girl, Martha. My niece and her husband are both secondary school teachers and in choosing what to call their daughter, they automatically qualified out scores of potential names, associating them with mostly negative characteristics of their female pupils. Of course, they rationally knew that choosing Olivia or Daisy wouldn't result in a mirror of the behaviour that bully Daisy terrorised fellow pupils with or that resulted in Olivia's expulsion. However, there was a bias at play and they selected a name with which they had no negative previous association.

Daniel Kahneman, a psychologist specialising in behavioural economics, judgement and decision-making, gives descriptions for many biases in his book *Thinking Fast and Slow*.[17] He suggests that we all engage in a number of fallacies and systematic errors, so if we want to make better decisions, we should identify the biases affecting our thoughts and consciously seek workarounds. Although Kahneman and his friend, the late Amos Tversky,[18] were psychologists, they were trailblazers in behavioural economics, paving the way for neuroeconomics which forms the basis of many of the neuroscience studies around decision-making.

Behavioural economics proposes a dual process theory – there are two systems or modes of thinking which we use to make decisions, intuition (System 1) and deliberation (System 2). The intuitive System 1 is more automatic and thought to operate outside of rational thinking. It results in faster decision-making that doesn't use working memory, but calls upon simple rules of thumb called heuristics, that can be applied generically for quick effect. System 1 is thought to be responsible for processing information relating to the gains and losses of decision outcomes. System 2, on the other hand, is slow, rational, takes mental effort and is responsible for higher order thinking and analytical approaches to decision-making. System 2 deliberates on information relating to probability outcomes and helps to inhibit non-serving impulses and breaking habits. Researchers from the Universities of Plymouth and Toronto reviewed criticisms of dual process theories and concluded that rapid autonomous System 1 processes are assumed to yield default responses unless intervened on by System 2 higher order reasoning processes.[19] The difference is that System 2 processing supports hypothetical thinking and relies heavily on working memory.

While Kahneman often discusses that biases can trip us up, German researchers found in research across decisions in businesses, healthcare and legal institutions that ignoring part of the information can lead to more accurate judgements than weighing up and considering all the information.[20] Inevitably shortcuts must exist in organisations as the landscape is constantly changing, so applying heuristic rules of thumb depends on the environment and situation at the time.

Always applying rational, algorithmic models across the board just won't work if an organisation is to be successful in an ever-changing world.

Cognitive biases and heuristics are useful for coaches to discuss with clients so they can build awareness of those they use. They can limit you and, like habits, may not serve you well – without you even realising it. They can also get you to a point of decision, rather than prevaricating and becoming stressed, which can be useful for the brain to stop decision fatigue and depletion.

Making better decisions

Fortunately, there are many strategies to help you make decisions that serve you better and to increase your confidence. Learn to listen to your intuitions, to take pause and hold your impulses and to consciously weigh up options to enable your rational brain to unify with emotional signals that come up. Coaching conversations can help you to become aware of biases at play and decision habits you may have become ingrained in without even realising it. Not deciding is frustrating for the brain.

According to the Zeigarnik effect, you are much more likely to recall uncompleted tasks than those you have completed. This effect originated over ninety years ago in a 1927 study by Russian psychologist Bluma Zeigarnik when she asked participants to complete a set of tasks.[21] During some of the tasks, participants were interrupted before they could finish. When asked later about the tasks, they recalled the tasks during which they were interrupted at a much higher rate than those they were able to complete. It seems your brain has a need to finish what it starts and if tasks are left incomplete, or goals unmet, thoughts pop up to remind you that there is still work to be done.

The Zeigarnik effect may explain why failures to act are easier to recall and more available as sources of regret. A study from a team at Cornell and Northwestern Universities found that people think about their biggest regrets of inaction more frequently than their biggest regrets about what they have actually done.[22] This research hit home for Laura, helping her to spring into a more proactive mindset. She recognised she had been stuck with indecision and more importantly, whether consciously or unconsciously, she was telling herself that her vacillation, avoidance and indecision was OK. She was adamant that any regrets she may have would not be from lack of action and therefore she needed to dust herself down and get on with her job – for herself, for her team and for the chance of success on the succession plan.

The end of the TV series cliff hanger is a classic example of the Zeigarnik effect in action. As coaches, we use it, maybe inadvertently, as well. Every time we open up a new avenue of thinking through our questioning, we pave the way for exploration in our client's mind. They may commit to think about trying out something new or have a eureka moment during discussion and are ready to explore further with high curiosity in the next session. Thoughts that were opened up simmer under the surface in the intervening period, not completed or concluded yet in the client's mind. Your conscious mind is looking for a decision about next steps, what to do with each open possibility.

Staying true to your values helps your decision-making

Many years ago, I was engaged by a client to lead the human resources team through their divestment and integration to a large multinational. Fortunately, the acquiring giant measured the deal's success by retention of all the staff moving over, so there would be no automatic job losses, just changes, in the process. Even so, there was a challenging time ahead to make sure the employees were matched to suitable roles in the new company and made the transition as pain-free as possible. The Human Resources Director invited me to speak at a board meeting some way into my contract and it was immediately apparent in the room that something was off centre. It transpired that a number of the senior team had been plotting the possibility of them negotiating imminent exit deals with large packages, bringing stock vesting dates forward so they could realise the cash benefits early and not move over to the acquiring company. On asking my opinion as to how to progress their plan, it was one of a rare few occasions when I was lost for words. In that moment, it appeared to me that the leadership of the organisation wanted to bail out with handsome pickings when the going got tough, leaving their loyal staff to work through the inevitable angst of being acquired. My recovery was helped by flexing my coaching muscles and asking questions. Whether or not my questioning gave any of the board members pause, I don't know. The contributing members were adamant to stick with the plan and a divide appeared in the room, with two heads of departments storming out, 'wanting no part of this'.

What happened in the room that day knocked me for six. Sometimes it takes a strong experience to validate what matters to you most. The behaviour and attitude of those four senior leaders flew in the face of two of my most treasured values – those of fairness and equality. Despite committing to all staff that the whole company was 'in it together', the team culture would remain and build as a growing but autonomous unit in the new company and that their doors were open at all times, on that day in the board meeting those doors had slammed shut. They broke their promise.

When working with my clients no matter what the presenting issues, we discuss personal values. Once you can express what is most important to you, whether it is achievement, family, spirituality, fairness, wisdom, service or wealth it brings clarity and congruence to your decision-making. Sometimes, like my example in the board room, it takes an act of great impact and repugnance that contravenes your values to the core to enable you to know for sure what matters in your life. Applying the Zeigarnik effect and not leaving you wondering, I publicly told the board on that day that I would be happy to help coach them personally through the transition to help them lead the troops and honour their promises, but I would not support them in their exit plan endeavours. In the end I wasn't fired, all the executives moved across to the new company, and two of the four 'moved on' to further their careers within the first six months. I still work with the Human Resources Director and one of the original board members.

Use your imagination to curb impulsive decisions

We often make trade-offs between immediate temptations and future payoffs. University of California, Berkeley researchers say that people who tend to give up immediate temptations and choose more patiently by using their imagination in those decisions may put themselves in a better position for the future.[23] Hitting the snooze button on the alarm clock might mean skipping breakfast and putting money into savings might mean being able to afford your dream wedding. Using imagination can promote patience by sidestepping exerting willpower over impulses by actually changing the impulses themselves. For instance, when the alarm rings to wake you up, imagining feeling energised and ready for the day after your favourite poached eggs and spinach may increase the desire to get out of bed, so diminishing the desire to press the snooze button. Laura and I discussed the impulsive temptations that had emerged through finding out she wasn't Paul's automatically selected successor and I encouraged her to use her imagination to reframe. Examples we brainstormed included imagining a refreshed Laura returning to work after a week's holiday rather than feeling she was missing out on a week of 'promoting herself' to her boss; imagining proactively being collaborative with Gail and Richard, gaining some new work insights and building trust, rather than her impulse to consider them 'competition' and the feeling that entailed; and imagining sharing a risky decision she took with her boss Paul asking for feedback, rather than just avoiding any decision, falling behind and maybe coming to Paul's attention for the wrong reasons.

Imagining future positive payoffs can help boost motivation and chances of goal achievement as well as encourage more patient decision-making. Chapter 6 provides a detailed look at the neuroscience of motivation, where making decisions for the future and curbing immediate temptations is pivotal for success.

Distance yourself from the problem

Stepping away from the situation as if you were an outside observer can help you make more rational decisions. University of Michigan researchers showed that when you think about a relationship conflict from a first-person perspective, you're less likely to use what they call 'wise reasoning' than when you think about it from a third-person perspective.[24] Wise reasoning includes strategies such as considering others' perspectives, thinking about different ways the scenario could unfold, and thinking about compromises. Become a fly on the wall looking in and act as though you were your best friend advising you. What would your best friend say to you, how would she coach you? Looking in as if it was your best friend facing the dilemma, what would you say to her and how would you coach her? We tend to make more objective decisions when we perceive the problems as belonging to someone else. This helped Laura to put her current dilemma in perspective, by hearing what her friend might advise. She also gained insight from our discussion of putting herself into Gail, Richard and Paul's shoes and viewing the world from their perspectives.

Learn to trust your gut

Some recent research on the complexity of making decisions based on gut feelings comes from Johns Hopkins Carey Business School in Baltimore.[25] Digging deeper into Kahneman's work, the study explores the idea that intuition can be a more useful tool than deliberate calculation in certain situations. The researchers propose that too much information can be just as misleading as a hunch in some cases, saying the human mind can be viewed as an adaptive toolbox. This toolbox consists of learned and evolved capacities such as memory, keeping track of movements, and social abilities such as imitation. Heuristics and what the researchers call 'building blocks' exploit these capacities to reach a quick conclusion. They discuss the tit-for-tat heuristic with the two building blocks of co-operating first, and then imitating the other person. Humans have the capacity to imitate with high precision from an early age and can use tit-for-tat almost effortlessly, whereas the question whether animals use tit-for-tat is still under debate. The researchers recommend that it's important to take intuitive decision-making one step further by recognising why people have developed such instincts and the best place to use them.

The power of sleep

You may have noticed an upsurge in discussions about sleep in the press and the shelves of your local book-shop. It appears that maybe, just maybe the message is getting out about how important sleep is not just for our bodies, but also our brains. As mentioned in Chapter 2 in discussing stress, a good rule of thumb is that what is good for your body is good for your brain and getting a good night's sleep, every night, is a great start in cutting your brain some slack.

Sleep continues to be of great interest to scientists looking to discover why it is so important for human survival. Sleep is not an inactive state, rather a period of time when strengthening and rejuvenation takes place. Just as it matters for your body to restore and regenerate, to grow muscle, repair tissue and to synthesise hormones, sleep also matters for your brain and for optimal cognitive functioning. It is required for storage of memories and lack of it impedes decision-making, concentration, reasoning and focus.

Business people, particularly senior leaders are sleep deprived, no doubt about it. A client once told me that it was almost a badge of honour to be at work before 6am, the job required it and he couldn't understand how on earth anyone managed to deliver in less than twelve hours a day. Like Laura, he often said 'you're a long time dead'. He was right. His sleep deprivation guaranteed he would get to that dead state earlier. Sleep has become associated with laziness. People perceive they are valuable if they are busy and a way of showing that is by proclaiming how little sleep they're getting.

Within the brain, sleep enhances your ability to learn, make memories, and choices. It fine-tunes your emotions, boosts your immune system, balances your metabolism, and regulates your appetite. Sleep allows the brain to sift through memories, forgetting certain things so as to remember what's important. One way it may do this is by 'pruning away' or 'scaling down' unwanted connections in the brain. This is thought to counterbalance the overall strengthening of connections that occurs during learning when we are awake. By pruning away excess connections, sleep effectively 'cleans the slate' so we can learn again the next day.

Getting enough sleep is important for attention and learning during your waking hours. When you are sleep deprived, you can't focus on large amounts of information or maintain attention for long periods. Your reaction times are slowed, and you are less likely to be creative or discover an innovative solution when trying to solve a problem or make a decision.

Sleep is also needed to do a bit of brain 'housekeeping'. Researchers at the University of Rochester Medical Centre have now started to unlock some of the mysteries around the mechanisms behind sleep benefits for the brain.[26] The team were interested in the lack of an equivalent lymphatic system in the brain and spinal cord to 'drain away' excess molecules such as proteins. The lymphatic system plays a critical role in the human immune system enabling the disposal of waste to the liver. The study found that cerebrospinal fluid, a clear liquid surrounding the brain and spinal cord, moves through the brain along a series of channels – managed by glial cells. It is as if the fluid acts as a 'sink' for waste and the brain actually can export molecules to the liver. Rodent studies show that the glia are the start of a transport network that ends up in the lymph nodes in the neck. The team termed this process the glymphatic system and reported it helps remove a toxic protein called beta-amyloid from brain tissue and their most recent research shows that sleep helps to clear these proteins. This has a huge implication for a number of neurological conditions such as Alzheimer's and other dementias which are characterised by an accumulation of proteins.

Last year, I attended a fascinating lecture by Matthew Walker, author of *Why we Sleep* and Director of the University of California, Berkeley's Sleep and Neuroimaging Lab.[27] He talked about our 'cultural sleep norms' being under assault on multiple fronts saying that midnight is no longer 'mid night'. For many of us, midnight is usually the time when we consider checking our email one last time. Making things worse, we don't catch up by sleeping any longer into the morning hours. He explained that our circadian biology, and the insatiable early-morning demands of a post-industrial way of life, denies us the sleep we vitally need.

If you're regularly sleeping less than seven hours a night, you're doing yourself a disservice as grave as that of regularly smoking or drinking to excess. And the trouble is that sleep-deprived people often don't recognise themselves as such.

Walker says that low level exhaustion becomes their accepted norm, or baseline. People fail to recognise how their perennial state of sleep deficiency has come to compromise their mental aptitude and physical vitality, including the slow accumulation of ill health. It is rare these declines are linked to lack of sleep.

A Carnegie Mellon University study showed that brief periods of what they called unconscious thought can improve decision-making.[28] The study gave participants information about cars and other consumer products while conducting brain scans. They manipulated participants into three groups – whether they consciously thought about this information, made an immediate decision about the products, or completed a challenging distractor task of memorising number sequences, to prevent them from consciously thinking about the decision information. When the participants were initially learning information about the cars and other items, the scans showed activation in the visual and prefrontal cortices, regions that are known to be responsible for learning and decision-making. Surprisingly, during the distractor task, both the visual and prefrontal cortices continued to be active, or were reactivated, even though the brain was consciously focused on the number task. Participants did not have any awareness that their brains were still working on the decision problem while they were engaged in the distractor number part of the experiment. Perhaps the most interesting result was that the amount of reactivation within the visual and prefrontal cortices during the distractor task predicted the degree to which participants made better decisions, such as picking the best car in the set compared to the other groups. Brief periods of unconscious thought were shown to improve decision-making compared with making an immediate decision.

There is something to be said for sleeping on it as Laura strongly advocated to me. She felt her usual decision style even under pressure was to take some pause, consciously weigh up options and for important decisions, if time allowed, she liked to sleep on it and allow her thoughts to 'land' overnight.

Challenge your biases

Until Roger Bannister achieved it in 1954, many believed that running the four-minute mile was a physiological barrier that no one could break, in much the same way as pilots had once regarded the sound barrier.[29] Bannister, a medical student, didn't see why such a barrier should exist. After all, a mile is simply an arbitrary measure that became a standard during the reign of Elizabeth I, and a minute is a sixtieth of an hour because the Babylonians, who originated the measurement, used a base-60 counting system. On 6 May 1954, Bannister ran a mile in 3 minutes 59.4 seconds, beating a record that had stood for the previous nine years.

Interestingly, Australian runner John Landy also ran a mile in under four minutes just forty-six days later and, by the end of 1956, no less than eight other athletes had achieved the same feat, proving that the barrier was never physiological but maybe was psychological.

We know from Kahneman's work that to survive in our complex world we rely on a number of heuristics, shortcuts or efficient rules of thumb, rather than just using extensive algorithmic processing.[30] These heuristics may also become barriers to your success unless you can bring them into your conscious awareness and do something about them. A coach can uncover a multitude of biases just through asking questions, noticing patterns and offering you a language for what they are hearing. To reduce biases, it is important to accept and acknowledge them and have an open mind to changing them. Like habits (Chapter 1) your bias neural pathways could be deeply embedded, so it will take energy to consciously train your mind to pause, check in and take another neural track. Being open to your emotional reactions and having the motivation to challenge them, like Bannister did all those years ago, is a route to changing them.

The frame effect is an interesting area to show the importance of being open to your emotional responses. Researchers from the Wellcome Department of Imaging Neuroscience at University College London looked at the neurobiological basis for the framing effect influencing decision-making.[31] They used gain frame trials as an amount of money retained from the starting amount such as keeping £20 of the £50, with the loss frame trials as an amount of money lost from the starting amount like losing £30 of the £50.

Increased activation in the amygdala was associated with a participant's tendency to be risk averse in the gain frame and risk seeking in the loss frame, supporting the hypothesis that the framing effect is driven by heuristics underwritten by an emotional system. The researchers suggest that the frame's impact on complex decision-making supports the role of the amygdala in decision-making.

Their findings suggest a model in which the framing bias can be a shortcut where you bring a broad range of additional emotional information into the decision process. In evolutionary terms, this mechanism may have given us a strong advantage, because such cues may bring useful or even critical information. Neglecting or ignoring the possibly unconscious cues may well impede the best decisions to be made in a variety of environments.

Sunk cost bias and meditation

A study from Insead and Wharton School of business found that mindfulness helped counteract deep-rooted tendencies of sunk cost bias and led to better decision-making.[32] Sunk costs are costs that you can't recover – something that you already spent and that you won't get back, regardless of future outcomes. It's like that gym contract you signed, whether you work out there or not, the money is gone and there's no way to get it back.

The researchers found that just fifteen minutes of mindfulness resulted in better decisions by considering the information available in the present moment, which led to more positive outcomes in the future. Using mindfulness could give various regions of your striatum and prefrontal cortex time to relay the true neuroeconomic costs of a decision and help you make smarter choices. Meditation reduces focus on the past and the future leading to less negative emotions, which in turn facilitates the ability to let go of the sunk cost. Mindful decision-making can derail impulsive, habitual and even addictive patterns of behaviour to help you make decisions more in keeping with your long-term interests and well-being.

Sunk cost bias can manifest itself in many ways and despite the implication from its title, it doesn't need to be about money. Staying in a bad relationship because you have invested so many years in it, eating all the food you order in the restaurant even if full, as you have paid for it, finishing an unenjoyable book or box set to the end just because you started it are all examples of this bias. For Laura, it was about being worried about 'competition' and not putting effort in because she had not been successful in her application for the top job anyway. Recognising that this bias was manifesting in her behaviours in this manner was useful and a catalyst for change.

Conflicting information and decisions

A University of Kentucky, Yale School of Medicine and Australian National University team used a simple gambling experiment to compare the two conditions of ambiguity and conflict.[33] They found that participants were more conflict averse than ambiguity averse and that ambiguity aversion did not correlate with conflict aversion; in other words, they were not a predictor of each other. The medial prefrontal cortex was more active when levels of ambiguity were high and activation in the ventral striatum correlated with conflict level and conflict aversion. These results contradict the hypothesis that a largely overlapping set of brain regions is involved in the processing of ambiguity and conflict and suggest that the ventromedial prefrontal cortex and striatum may play a more separate role in the processing of either ambiguity or conflict. This will be an interesting area for further research but whether overlapping or distinct functioning, clearly our ability to make effective decisions is affected by the quality of information we receive.

Team conflict

Laura particularly struggled to reach a decision if there were too many options on the table and receiving her balanced MiND profile feedback where she could typically see value in most viewpoints had helped her to understand her behaviours (see Appendix for details of the MiND tool). Recently in team meetings, if she was asked to voice an opinion, she tended to vacillate and add nothing new. Laura suggested to her manager Paul that it would be useful for me to facilitate a MiND workshop for the whole team which he agreed to wholeheartedly. Paul was a collaborative manager, ran regular team meetings and felt such a team workshop would add value not just for each team member but for the team as a whole.

The workshop was a revelation to many team members and it gave Laura information to help understand her responses within the team. Their profiles revealed a heterogeneous group with multiple brain dominance strengths and preferences. Seeing her profile in comparison to her colleagues' helped Laura to understand how their motivations and energy differed in their natural approach to work tasks and decision-making. The team discussion that emanated focused on respecting each other's strengths, enabling everyone to have a voice at team meetings and using the MiND model as a problem-solving tool and a way to understand customers better. To do this the whole group 'steps into' the different dominances of Reasoning, Spontaneous, Emotional and Specific styles and imagines how a person with a strong neurological dominance in that area would solve the problem or make a decision. After the workshop, Laura's energy for her personal development was magnified one hundredfold. It was like turning the volume up on our conversations and she shared that the atmosphere in the office was transformed, buzzy, energetic, full of banter and fun. And most importantly she felt she had a part to play on Paul's senior team, her contributions were respected and Gail and Richard were allies, not 'competitors'.

Understanding the neuroscience behind making a decision can be helpful when you are trying out new behaviours and challenging your automatic habits. When you reach a fork in the road or get stuck in a pothole and need to make the right decision for your long-term health, well-being or progress then using the brain science behind decision-making is a useful tool.

CHAPTER SIX

The Motivation Mission

'The older you get, the more fragile you understand
life to be. I think that's good motivation for getting
out of bed joyfully each day.'
  – Julia Roberts, actress

N ew terms abound in the business world and there
are two I discovered recently. Brownout has been
used to mean low level employee malaise by US coaching company Corporate Balance Concepts.[1] In the
industrial world a blackout means that all power has
been lost whereas a brownout is when the plant runs
on lower wattage. Transposing that definition into the
working world, it's a way to describe people not firing

on all cylinders, where their spark is disappearing, they are becoming disenchanted, demotivated and disengaged. The company report around 40% of the 1,000 executives they surveyed suffer from some sort of brownout which could then spiral into burnout which an estimated 5% of respondents reported. Shocking figures that certainly should give organisations food for thought.

Presenteeism was another area of interest I heard about at the Scottish CIPD Conference[2] in 2016 during a keynote speech by Professor Cary Cooper,[3] who many years ago I was lucky to have as a lecturer during my first degree at Manchester University. Describing presenteeism as when employees come to work unwell, unable to perform at their best, he revealed estimated cost figures for the UK employers at £15.1 billion per year.[4] In the presentation, he discussed how stress and workplace conflict affect employee well-being and company cultures, how studies show only 35% of employees are truly present and healthy and told of the consequences of physical health issues to employee engagement, productivity and motivation.[5]

No matter what descriptor is used, brownout, presenteeism, employee engagement and motivation are becoming real issues for organisations and ones where MyBrain Practitioners are increasingly focusing their work. There are many different reasons for this – changing roles, flexible working, cultures of long hours, continuous change and restructuring, mergers, acquisitions and increasing insecurity and uncertainty.

The revolution of technology in our lives and work, while representing progress in terms of automation of tedious tasks, has also resulted in increasing complexity of roles and added in 24/7 contact which some people just can't disconnect from. Against this backdrop, it's no wonder that some people feel disenfranchised and lose motivation.

Behavioural psychologists often define work into two groupings, algorithmic or heuristic. The former is routine based where you can follow instructions to complete the work. The latter requires thinking, there is no algorithm or instruction book, you need to try out different options and come up with novel solutions. As well as being absorbed into the technological interface at work, much of the routine algorithmic work is now outsourced to countries where it is cheaper to get it done. As a result, organisations are now requiring their employees to engage in an increasing amount of heuristic work much of which depends on an internal drive – or intrinsic motivation. This type of motivation is all about engaging in an activity because you find it enjoyable, interesting and rewarding. Extrinsic motivation on the other hand is when you engage in an activity in order to gain a reward or avoid a punishment. Many of the pay for performance or bonus systems embedded in organisations' performance management systems are an attempt to extrinsically motivate their employees to work hard to achieve their objectives at an optimal level, but they are not always fit for purpose in a world of heuristic work. For me, writing this book on a daily basis is driven by intrinsic motivation as I enjoy exploring new research, getting lost in the writing and feel totally energised by finding novel connections I can use in my work. There are also elements of extrinsic motivation such as the drive to have the book published and available for a conference I am speaking at later in the year and the avenues it may open up for networking and future collaboration.

Dan Pink's fascinating book *Drive: The Surprising Truth About What Motivates Us* pulls together a rich merging of research from psychology, sociology, neuroscience and other disciplines.[6] He talks about changing times but lagging companies, which aren't progressing in applying what is now known from science about motivation. He suggests that the world has moved on from 'motivation 2.0' which is about rewards and punishments to a world of 'motivation 3.0' which is all about inherent satisfaction in the work itself. Some situations such as routine tedious work may still benefit from the incentives of 2.0 but for newer, creative, thinking-based work, they may have a limiting, even destructive effect. Motivation 3.0 has three elements – autonomy, where people want to have control over their work, mastery (or competence), where people want to get better at what they do and/or relatedness (or purpose) where people want to connect and be part of something bigger than they are.

These areas were originally proposed as part of Self-Determination Theory many years ago by psychologists Edward Deci and Richard Ryan from the Australian Catholic University in Sydney and University of Rochester in the US respectively.[7] The theory has been refined and researched by many academics since and this chapter explores some of the underpinning neuroscience around what Pink calls motivation 3.0. Motivation is a large subject and links to further areas of curiosity, play and flow which will also be discussed.

MARK'S STORY

Mark worked for a leading pharmaceutical company in HR as part of the European Corporate University – the learning and development department. He travelled extensively delivering leadership training sessions and supporting the business heads across the EMEA (Europe, Middle East and Africa) region, designing solutions for their leadership development needs. He had been headhunted into the company five years ago and was part of the high-potential pool, identified as future senior leadership. He was in discussion with his manager about taking additional responsibility for leadership training within the Asia Pacific region when his colleague in Sydney went on maternity leave. He believed this year-long opportunity would place him in a good position for suggesting a permanent global position for leadership development in the future – his ideal career ambition. In addition, he was studying for an MBA and was mentored by the UK country manager. Life was good at work for high achiever Mark and he believed he was on the cusp of putting a tangible plan in place to realise his career goals.

As a leading organisation, with state of the art, sophisticated HR programmes, Mark's company was able to mobilise its succession plan quickly when the business required. A shock vacancy due to illness of a board member triggered a series of promotions across the company, including Mark's manager, and he found himself with a job offer of Director for the Europe, Middle East and Africa (EMEA) Corporate University, reporting to the Global Corporate University Vice President. When telling me his story he recalls feeling 'super pumped' and flattered, believing this change in his career aspirations could be a fast track to even bigger and better roles in the future. He had a highly motivating salary negotiation session with the Head of EMEA Human Resources and an encouraging conversation with his mentor and took the role with little hesitation.

Three months later Mark sought my help. He reported he had lost his mojo and believed he had made a monumental mistake, even considering looking for a new opportunity in a different company. His replacement in his old role was doing really well and was sharing the Asia Pac maternity cover with a US colleague. Mark was clear that there was no going back but was also forthright in his view that the lead Corporate University job held little for him. Specific examples he cited that he 'hated' were the management of the P&L, continuous budget justifications, amassing ROI data, less client interface and touching the real business – the employees in the training room. He still saw his mentor on a monthly basis who had recommended coaching as a potential way to gain perspective and talk through options.

Mark was stuck, demotivated, fed up, yet he had the world at his feet.

The brainy bit

Human beings are driven to undertake activities that reward you or remove you from threat or punishment. This drive towards or away from something will reinforce behaviours so you are motivated to take the same action again. At a fundamental level, your brain treats survival needs as rewarding and motivates you to fulfil those needs. Most neuroscientific research around this area has been designed predominantly around the neural mechanisms of extrinsic motivation, incentives or rewards. This chapter explores intrinsic motivation in some detail, showing some similarities and distinctions.

Research on incentive motivation involves the 'reward system' in the brain whose main neurotransmitter is dopamine. These reward circuits include the dopamine rich neurons of the ventral tegmental area, the nucleus accumbens and part of the prefrontal cortex and uses the mesolimbic pathway as a major transit route.

A collaboration between the Universities of Rutgers and California, Los Angeles showed that these circuits are activated by performance feedback, even in the absence of extrinsic rewards, registering the feedback signals and dopamine elicited when many activities are carried out.[8] They found that brain activity is modulated by different contexts such as whether feedback is about goal achievement, whether people are motivated to perform well relative to their peers and whether learners are motivated towards feedback of an informative nature, moving towards a choice in the future or an evaluative nature, learning from something already performed. The types of activities that people find intrinsically motivating are stretching but achievable, with clear goals that are not too far away in the future and those with immediate feedback.

Extrinsic motivation can backfire

In his book *Payoff: The Hidden Logic That Shapes our Motivations*, Dan Ariely, Professor of Psychology and Behavioural Economics at Duke University, shares some fascinating studies showing how rewards can impact intrinsic motivation.[9] This is referred to in different literature as the undermining effect, the motivation crowding-out effect or the over justification effect.

In an experiment at a semiconductor factory, employees were given a number of different incentives at the beginning of each day to try to increase productivity. One group was told that if they hit their targets, they would receive a bonus of $30, another group were promised pizza vouchers, and a third group were told they would receive a note of thanks from the CEO. Employees in a fourth group were not offered any incentive. Productivity only increased 4.9% for the bonus incentive group. The pizza voucher was the winner, with a 6.7% increase in productivity and the CEO's thanks interestingly came a close second at 6.6%. On the second day, the cash incentivised group performed 13.2% worse than employees who were not offered any incentive. Although this difference decreased to 6.5% on the third day the effect of the cash bonus could be summarised as both increasing costs for the company while actually decreasing employee productivity.

Research from Germany and Japan studied the neural basis of this undermining effect using fMRI scanning.[10] Volunteers were randomly divided into two groups – a reward group that received 200 yen (around £1.50) for each successful performance and a control group that received no payments. Participants played a stopwatch task where they pressed a button with their thumb within 50 milliseconds of the 5 second mark, gaining a point when they succeeded. During an initial scanning session, participants in both groups showed greater activity in the midbrain and caudate head (bilateral anterior striatum) when given success feedback relative to failure feedback. The success feedback involved reward network activation whether the group was given a monetary reward or not. During the first session, the lateral prefrontal cortex in the reward group showed significantly larger activation than that in the control group, suggesting that participants in the reward group prepared for the stopwatch task more actively than those in the control group when they saw a task cue. A 3-minute free choice period was then allowed where participants were left alone in a quiet room to do anything they wanted. People in the reward group were less likely to voluntarily engage with the task during that period and this 'behavioural undermining' of intrinsic motivation was accompanied by reduced activity in the caudate and midbrain during a second scanning session when monetary rewards were no longer given to the reward group. In contrast the unrewarded group stayed at the previous levels of activation. The lateral prefrontal cortex activation also became significantly smaller in the reward group than in the control group. The researchers concluded that this difference in activity between the groups is consistent with the idea that your dopaminergic value system responds to cues that signal progress on the task when you are intrinsically motivated. Specifically, when performance-based reward is no longer promised, people do not feel much value in succeeding in the task, as indicated by the dramatic decreases in the activation of the striatum and midbrain in response to the success feedback. They are also not motivated to show engagement in facing the task, as indicated by the decrease in the lateral prefrontal cortex activation in response to the task cue.

Mark found our discussion around extrinsic and intrinsic motivational differences interesting. He shared with me that he was a member of the team who influenced a change to the 'pay for perfomance' system in his organisation two years prior after some vociferous

feedback on the employee opinion survey. Traditionally the company had offered bonuses for performance
against individual, team and company objectives and
had a forced ranking system to decide top and bottom
performers and everything in between. He explained
how demotivating the system was and how it was a
major factor in the previously declining engagement
results.

Autonomy

Self-Determination Theory says that humans have
three basic psychological needs – autonomy, competence (or mastery) and relatedness (or purpose). A
sense of control and the opportunity to choose is a way
to offer more self-determination over your behaviour
so your self-perceived autonomy can therefore help
your intrinsic motivation. The theory also suggests that
support for autonomy and intrinsic motivation will
lead to other adaptive outcomes including increased
performance, learning and even better health, so work-
places would do well to look at what they can do in
this area.

A collaboration of researchers from the Universities
of Reading, Rochester, Tokyo and California Institute of Technology looked at 'self-determined choice'
using brain scans.[11] Participants played a game using
a stopwatch with either one they selected themselves
or one they were given. The results showed a better
performance from the group who had selected the

stopwatch even though choosing was irrelevant to how difficult the task was. Feedback about failing on the task, compared with feedback received when successful resulted in a drop in the ventromedial prefrontal cortex activity for the forced choice stopwatch but not for the participants who had chosen the stopwatch. The resilience to failure of the ventromedial prefrontal cortex was significantly linked to the increased performance showing its critical role in autonomy on results.

Researchers at Zhejiang University studied autonomy between time-estimation tasks with two groups again either self-selecting the task or given no choice.[12] EEG waves were recorded from more than sixty sites on the scalp to explore what the researchers called the 'electrophysiological signature' of autonomy and also anticipation. The EEG amplitude analysis showed that the choice option increased the motivational significance of the performance feedback and a heightened expectancy may be the result of an enhanced sense of control when choice was available. Participants also perceived a greater loss when they failed on a chosen task. An interesting result showed that increased effort didn't actually result in higher performance, which doesn't hold with Self-Determination Theory. The researchers recognise that choice may lead to more care and thinking around performance which could result in worry over its outcome. Overall however, the study does help to demonstrate the important role of autonomy in the workplace and the researchers state that no matter how trivial and inconsequential the choice is, offering it might make a difference to intrinsic motivation.

Despite moving to a more senior position with wider reach and influence, Mark felt his autonomy was greatly diminished and believed it to be a major factor in him losing his spark. He told me that intellectually he knew it wasn't as bad as he made it out to be, but as he equated the role with reporting numbers and rationalising data, he felt he was always required to give an explanation or justify himself. His demotivation had catalysed a generalisation distortion and a negative perspective previously not experienced, which concerned him and he recognised as a potential downward spiral (see Chapter 3 on negative thinking for an explanation).

Competence (mastery)

According to Self-Determination Theory, people are naturally inclined to interact with the environment in ways that promote learning and mastery. Verbal feedback which emphasises your competence in a task is likely to enhance your perceived competence so leading to increased intrinsic motivation.

German and British research used picture puzzles where volunteers were shown a relatively unknown modern art picture that was displayed twice.[13] One was the original version, the other one showed up to four small differences. Different groups were given either verbal performance feedback or a cash reward or no reinforcement. The researchers wanted to see what happened when different types of rewards were applied and then taken away. All groups reported that the 'spot the difference' exercise was fun and they were engaged. Activation in the anterior striatum and midbrain was higher after verbal rewards than in the control group suggesting that people have a higher subjective value for succeeding after verbal reinforcement. Performance increased significantly over time only in the verbal rewards group even when positive feedback was withdrawn, although the actual increase wasn't significantly larger than in the other groups. The researchers speculate even that low intrinsic motivation was increased by verbal reinforcement but recognise that data was not strong and that more research is needed.

Mark's new role involved reporting to a manager on the west coast of the US, eight hours behind the UK. While his manager was approachable and supportive, Mark was finding the time difference challenging to even keep him up to date on key projects, let alone ask him for feedback to help him progress.

The insula may be a unique neural element of intrinsic motivation

The striatum is widely accepted as a brain region that plays an important role in the generation of motivated behaviour whereas the insula is well known for the processing of emotion and feeling. However, University of Iowa researchers found insula involvement for intrinsic motivation when they compared patterns of brain activity when participants imagined acting for intrinsic and extrinsic reasons.[14] Examples included interesting or enjoyable activities such as working on the computer out of curiosity, participating in a fun project, writing an enjoyable paper for the intrinsic condition. Different incentive-based scenarios were posed for the extrinsic scanning such as writing an extra credit paper, participating in a money-making project or working on the computer for bonus points. Results showed that the right insular cortex was more activated in the intrinsic motivation condition while the extrinsic motivation condition showed greater neural activity of the right posterior cingulate cortex, which is a brain region of the valuation system, than in the intrinsic motivation condition. An interesting conclusion was that 'engagement decisions based on intrinsic motivation are more determined by weighing up the presence of spontaneous self-satisfactions such as interest and enjoyment while engagement decisions based on extrinsic motivation are more determined by weighing stored perceived values as to whether the incentive is attractive enough to warrant action'.

Mark described himself to be hugely motivated by learning new things and performing them to a world class level. The trouble was that as time went on he was discovering he wasn't enjoying his role, resulting in diminishing interest, application, curiosity and learning. By all accounts, he was doing a good job, as he was a high achiever and always strived for high performance, but he wasn't motivated to make mastery (of his role) part of his vocabulary.

Prediction errors

A Cambridge University study showed that dopamine signals can be useful for learning long chains of events.[15] You constantly update your goals through a dopamine-driven process called reward prediction errors. These signals are the difference between the reward that is actually delivered and the reward that is predicted to be delivered. Processing of prediction errors rather than full information about something saves neuronal information processing. Reward-predicting stimuli are conditioned rewards and have similar effects as real rewards on approach behaviour and learning. Dopamine neurons treat reward predictors and real rewards in a similar way, as events that are valuable for the individual. The dopamine neurons provide information about past and future rewards that is helpful for motivation, learning, planning and decision-making. You choose actions based on predictions of good or bad outcomes and if the result is better than expected then the reward system drives you to repeat the same behaviour next time.

This can be a useful discussion to have with clients. If a task repeatedly doesn't go as well as expected then you are likely to predict a continuing declining result, becoming less rewarded – and so it goes on. For high achiever Mark, always hungry to learn and grow, a decrease in his motivation was resulting in a lower prediction of the reward that the learning could offer him. He found himself slowly closing doors to that opportunity, and our conversations helped to keep them ajar then open them again.

Purpose (relatedness)

People are motivated and demotivated by different things. The strength of connections to the parts of the brain that connect to the reward systems depends on both genetics and experience. Some people will be motivated by social reward and enjoy working in a team, while others will be more motivated by achieve-ment so will work hard for a high mark. Some people self-identify with their role more than others so will be more sensitive to job titles while others will be affected by other people's opinions and may be nervous about a new manager.

One of the best ways to connect with your intrinsic motivation is to know why you are doing something. What difference do you want to make, what legacy do you want to leave? Being able to link your daily activities, including the more boring, repetitive ones, to your higher purpose helps keep up your intrinsic motivation, driving you to achieve your goals.

A frequently viewed TED talk is Simon Sinek's *How Great Leaders Inspire Action* which I encourage all my clients to watch.[16] He talks eloquently about discovering your 'Why', your purpose, cause or belief. In the talk, and his book *Start with Why*[17], he introduces the golden circle, a graphic consisting of three concentric rings labelled from the outside in 'what', 'how' and 'why'.

What's are easy to identify. Every organisation knows what they do and everyone is easily able to describe the products or services a company sells or the job they have. The how is about what makes the company and its people unique, different or better. Some companies know how they do what they do, but it is incredible how many don't. Few people or companies can clearly articulate why they do what they do. It isn't about profit, that is a result. Why is all about your purpose, cause or belief. Sinek asks, why does your company exist? Why do you get out of bed in the morning, and why should anyone care?

When most organisations or people think, act or communicate they do so from the outside in, from what to why, going from the tangible to the intangible. Sinek says we say what we do, sometimes how we do it, but rarely say why we do what we do.

Taking Apple as an example he demonstrates his golden circle by sharing a possible marketing message from the outside in – 'we make great computers, they are beautifully designed, simple to use and user friendly, want to buy one?' This is the classic what, how – we say what we do, we say how we are different from our competitors and we expect some sort of action, like a sale. In reality though, Apple starts its communications in the inner circle and Sinek gives an expansion of how that sounds, starting with the Why: 'Everything we do, we believe in challenging the status quo. We believe in thinking differently. The way we challenge the status quo is by making our products beautifully designed, simple to use and user friendly. We just happen to make great computers. Want to buy one?' It's all about connecting with people who believe what you believe and Sinek shares that this fundamental shift is grounded in the 'tenets of biology', loosely correlating the neocortex with his golden circle's 'what level' with the 'how' and 'why' emanating from the limbic brain. When we communicate from the outside in elaborating features and benefits, facts and figures, it doesn't drive behaviours. When we do it from the inside out, we are talking directly to the part of the brain that drives behaviour, connection and change.

Goals are about the pursuit of a result but intrinsic motivation to achieve truly comes from an understanding of your purpose and your why. Sinek says that success comes when you are in clear pursuit of why you want what you want. Your achievements en route serve as the milestones to tell you that you are on the right path. In the course of building a career you become more confident in what you do and more expert in how you do it. But if you don't have the why then motivation for the what can slip through your fingers and possibly become misaligned.

Mark and I shared a rich conversation around discovering his why. He realised he had allowed himself to get caught in a negative spiral around his new role, completely losing perspective about what he wanted in his life and what an amazing job it was. We discussed multiple scenarios and his framing around each, from leaving the company, to making the best of a bad job, to looking at the role from a different angle. Through deep reflection Mark found he had allowed himself to become disengaged from his why, which he recommitted to during our sessions. He wrote in his journal that he wanted to be an enabler in the world, helping people to be the best they can be, giving them confidence, showing them the way to freedom and a life of their choice. He had been clear of a 100% match in his previous role as a facilitator of training, face to face with employees, however, he had not consciously done the work to ensure his new role was aligned to his why. Through a series of exercises, Mark looked at the elements of his new role and wrote a statement for how each helped to progress him even deeper into his why. Even running a training P&L for the Corporate University scored. He reframed that as him being responsible for ensuring that every employee across the organisation was allocated training dollars for their personal development. He also decided to become a mentor for younger members of staff to keep his contact going deep into the business rather than just at the executive level. He reported that in so doing, he knew he was congruent with his why, as he was happy, energised, and felt he was contributing across the business and was in flow.

Flow

Have you ever been completely absorbed in what you are doing, in the zone, where the task seems effortless? This is the experience of flow so called by Hungarian-American psychologist Mihaly Csikszentmihalyi after his adverse experiences as a prisoner during World War II and his consequent studies around contentment and happiness.[18]

He defines flow as 'A state in which people are so involved in an activity that nothing else seems to matter; the experience is so enjoyable that people will continue to do it even at great cost, for the sheer sake of doing it.'

Using PET scanning, researchers at the Karolinska Institutet, Sweden, found that people prone to experience intrinsically motivated flow states in their daily activities have greater dopamine D2-receptor availability in striatal regions, mainly driven by the dorsal striatum, with a significantly higher correlation in the putamen than in the ventral striatum.[19] This finding suggests that people's capacities for intrinsic motivation are associated with the number of receptor targets within the striatum for dopamine to act upon. The researchers also suggest that proneness to flow is related to personality dimensions that are under dopaminergic control and characterised by low impulsiveness, stable emotion, and positive affect.

Prefrontal cortex activity during both flow and boredom states when participants were playing the video game Tetris was studied by a University of Hokkaido team using functional near-infrared spectroscopy.[20] During flow, activity was significantly increased in the right and left ventrolateral prefrontal cortex. Activity tended to decrease when participants were bored. Consistent with the idea that intrinsically motivated states recruit central executive regions, flow activates the prefrontal cortex, and may therefore be associated with functions such as cognition, emotion, maintenance of internal goals, and reward processing.

Curiosity

Curiosity is a particular system of intrinsic motivation that drives you to learn something new. A study from the University of California, Davis found a link between brain mechanisms of the reward circuits and dopamine supporting extrinsic reward motivation and intrinsic curiosity.[21] Their research suggests that when your curiosity is piqued, changes in the brain enable you to learn not only about the subject at hand, but also incidental information, highlighting the importance to create a more effective broader learning experience.

Participants reviewed more than 100 questions, rating each in terms of how curious they were about the answer. During brain scanning, they then revisited 112 of the questions – half of which they were strongly curious about and half they found uninteresting. After a question, a completely unrelated photo of a face was introduced before the answer was revealed. The researchers then tested participants to see how well they could recall and retain both the trivia answers and the faces they had seen.

Findings showed that greater interest in a question would predict not only better memory for the answer but also for the unrelated face that had preceded it. A follow-up test twenty-four hours later found the same results – people could better remember a face if it had been preceded by an intriguing question. Midbrain regions were of interest, the left ventral tegmental area, bilateral nucleus accumbens and the dorsal striatum, suggesting that before the answer had appeared the brain's curiosity was already engaging the reward system through anticipation. Additionally, there was increased activity in the hippocampus during recall of incidental face stimuli during high curiosity questions and increased functional connectivity between the hippocampus and ventral tegmental area when participants anticipated answers to the trivia questions. In fact, the degree to which the hippocampus and reward pathways interacted could predict an individual's ability to remember the incidentally introduced faces. The researchers concluded that when people were highly curious to find out the answer to a question, they were better at learning that information. More surprising, however, was that once their curiosity was piqued, they showed better learning of completely unrelated information that they were exposed to but not necessarily curious about. People were also better able to retain that incidental information learned during a curious state across a 24-hour delay. When curiosity motivated learning, there was increased activity in the hippocampus, a brain region that is important for forming new memories, as well as increased interactions between the hippocampus and the reward circuit.

This research supports a greater learning experience from students participating in classroom training. As in Mark's department in his Corporate University, many learning and development professionals have been frustrated by the continual cuts in training budgets and the move to online learning, e-learning or a blended formula. While the argument is that an online course delivers the content in a focused way, there is so much missed by not engaging in conversations with others. At MyBrain International, we strongly believe in the power of discussion, hearing and telling stories and generating energy and ideas in the training environment. Learning isn't just about gaining answers, it is also about exploring, discovering and asking questions – thereby broadening out rather than narrowing down. Curiosity is a natural, human way to learn about new areas you may not have yet considered could be enhancing for you.

Enhanced memory from a curious state was again highlighted in a Dutch study into perpetual curiosity.[22] Blurred pictures of common items were shown to participants to induce curiosity before offering clear unambiguous versions.

The resolution of perceptual curiosity was associated with activity within the left caudate, putamen, and the nucleus accumbens, regions that we know comprise the core of the dopaminergic system related to reward processing. When the curiosity was resolved, participants showed activation in the hippocampus and enhanced incidental memory. The researchers suggest the results provide neural explanation for the classic psychological theory of curiosity, which states that curiosity is an aversive condition of increased arousal. The resolution of the curiosity is rewarding, facilitating memory and learning.

Play

Two of my best friends are brown and furry. Gordon and Charlie are seven-year-old chocolate labradors that are naughty, disobedient, cheeky, loyal and full of unconditional love. Come rain or shine, every morning begins with a run or a walk with the dogs. No matter how cold or wet the weather is, how tired or busy I am, I always enjoy being out with my boys in the morning and it gives me a great start to my day. Every day they make me smile, sometimes laugh out loud, with their antics, their sheer joy at being outside, sniffing, exploring and no holds barred playing. Think about the contrast between the play environment at a children's nursery and school, where play is slowly phased out of the mix. Nursery is often the first place a child goes for any length of time without their parents – and what an exciting place. Brightly coloured walls and smiling staff beckon in a messy and curious environment and children are encouraged to play and express themselves as they wish, just like Gordon and Charlie. Some get involved in everything while others choose to play quietly in a corner. Differences are acknowledged and children are encouraged to join in the hubbub of creativity as they learn and play at their own pace and in their own way.

That all changes when a child goes to school – at four years old in the UK – when even in the younger years the life becomes more ordered and routine. Children wear uniforms, they line up in playgrounds, they have their own desk, their books are organised neatly in their desks, the timetables are routine and lessons follow an agenda. The older a child gets, the less pictures they will see in books, the less time they will be able to run around the sports field, splash paint onto walls, play games and chat about their dreams.

This is not to say that routine, order and discipline are bad in themselves, just that an emphasis on these tends to restrict and inhibit the natural play instinct.

Stuart Brown, founder of the National Institute for Play and author of *Play: How it shapes the brain, opens the imagination and invigorates the soul,* has gathered thousands of case studies that he calls play histories.[23] He believes that remembering what play is all about and making it part of our daily lives are probably the most important factors in being a fulfilled human being, stating that 'the ability to play is critical not only to being happy, but also to sustaining social relationships and being a creative, innovative person'. We all start out playing naturally and from our play, we learn how the world works, how people interact, how exciting life can be. As we grow up even sports or hobbies can become routinised, competitive and full of instructions which can take the joy out of free play. In most cases play is a catalyst to new ideas, to productivity, to rebooting and finding some joy in the moment. It certainly isn't something we should feel guilty about as adults or believe it to be the opposite to work. Brown says that life without play can be a grinding, mechanical existence organised around doing things necessary for survival 'play is the stick that stirs the drink... (it) is the vital essence of life.... We are designed to find fulfilment and creative growth through play.'

Researchers at the University of Lethbridge in Canada reviewed multiple studies around the function of play in the development of the social brain in rats.[24] Rough-and-tumble play, or play fighting, is common in the young of many mammals. Research on play fighting among rats shows that subcortical structures mediate the motivation and behaviour of play, and the cortex mediates the effect of play with age to refine into social skills. The 'Motor-training Hypothesis' states that play during the juvenile period prepares the motor system of animals for engagement in adult behaviours whereas the 'Training-for-the-Unexpected Hypothesis' suggests that when animals play, they expose themselves to lots of different actions, many of which lead to unpredictable results. This then helps them to learn to cope in an unpredictable world. Play in the juvenile period may enhance skills in adulthood by dampening the fear of novel situations where the prefrontal cortex dampens the activity of the amygdala to prevent emotional over-reaction. Their findings strongly suggest that exposure to play directly and indirectly influences the development of the prefrontal cortex, therefore the dampening of amygdala activity could be the vehicle play uses to train animals to be more resilient in an unpredictable world. They conclude that for rats, the two hypotheses – training for the unexpected and extended motor training – converge into one process such that play trains animals to be resilient by modifying the neural circuitry that regulates emotional responses. Such cortical changes appear to mediate the effects of play on the refinement of social skills. As a result, rats that play as juveniles are more socially competent as adults.

It was as if Mark had won the lottery when I introduced the research around play to him. He became animated and energised, claiming that he now understood the real reason he wasn't feeling motivated about his new role, despite intellectually getting that it was a good career move. He told me it was because he was no longer having fun and he now recognised that the reason he enjoyed every second of his previous job was because it didn't feel like work. We discussed elements of his new role and how he could reframe them and shape them as energising, fun ways to spend his time.

Give yourself a dopamine boost when doing tedious tasks

Dopamine is stored abundantly in the nucleus accumbens which is also sensitive to other neurotransmitters like serotonin and endorphins. The trick is to identify ways to get a dopamine boost while you are doing the boring, repetitive tasks that lead to longer term payoff and reward. Keeping a list displayed in front of you and ticking off what you have achieved is helpful, listening to your favourite music, consciously connecting each task to your why or your goals. Extrinsic motivation can be useful for basic, repetitive tasks, promising yourself a coffee break after completing something, or a social media fix, but remember it can reduce intrinsic motivation, stopping creativity, forcing you to think short term and undermine your intrinsic motivation.

You don't have to resort to sex, drugs and rock 'n roll to give yourself a boost, although maybe a bit of consensual sex would be good. Healthy options include eating dopamine boosting foods such as bananas, leafy greens, oats and avocados. I love chef Tom Kerridge's cookery book *Dopamine Diet*, packed full of delicious recipes built around tyrosine-rich ingredients known to facilitate the release of dopamine.[24] L-tyrosine is an amino acid used to produce dopamine and noradrenaline. As it is commonly found in protein rich foods, which can make you feel fuller for longer, a common side effect is weight management – and Tom Kerridge with his massive 11 stone weight loss is testament to that, saying he has never felt hungry or deprived while following it. As discussed in Chapter 2 on stress, exercising outdoors, massage and meditation are also great ways to boost your levels.

Priming

Expecting to be upbeat and in a good mood all the time just isn't realistic. But learning what helps you to bust any negativity, demotivation or just a 'whatever' mood can help you take proactive moves. Ask yourself, what makes you feel good? Is it just asking your Alexa to shuffle a load of Take That songs, is it putting some space between your current mood and the state you want to elicit, by for instance going for a walk or watching a quick TED talk? Is it leafing through some favourite photos to remind yourself of happy times or reminding yourself of your why by standing in front of your vision board? If I feel a bit fatigued or depleted I tend to become distracted from the task at hand and sometimes it takes mental effort to get my motivation back on track. I use several strategies to help, such as going out for a 20-minute jog or if working from my home office I go into my lounge and look at my family photos. Some of the photos are formal shots, but most are funny group snaps from happy days shared. My mum died nine and my Dad four years ago, but I still talk to them, often 'face to face' with the collection of framed photos all over the shelves. Sometimes I feel a little melancholy but I always feel grateful as they were fantastic parents and I had a wonderful upbringing. I find it reboots me, reminds me that I come from good stock and know how they would be encouraging me if they were in the room with me. These personal strategies can help to elicit a more positive state for your motivation to return. What you are doing is priming yourself into a more motivated positive state as I do when I talk to my parents.

Priming refers to activating particular representations or associations in memory just before carrying out an action or task. For instance, I see photos of my parents, 'hear' their voices and feel encouraged and grateful.

A great example of priming comes from the world of illusionists. Derren Brown is a master at bringing people on stage and seemingly getting them to say what he predicted they would.[26]

In an episode of his TV series *Mind Control* he invited two employees of a London advertising agency to a secret location and appeared to influence them to come up with a name, a tagline, and a logo for a fictitious new taxidermy business.[27] They decided on the name 'Animal Heaven: The Best Place for Dead Animals' with a grizzly bear holding a lyre logo, zoo gates resembling the gates of heaven, and angel wings above the business name. Opening a sealed envelope that had been in front of them the whole time, Brown remarkably revealed an almost identical name, tagline, and logo 'Creature Heaven: Where the Best Dead Animals Go,' with a logo of a bear holding a lyre, angel wings, and pearly zoo gates.

The programme then backtracked to the taxi ride they had taken to travel to the film location, driving past London Zoo showing shot after shot of lyres, angel wings, pearly gates, and the words 'Creature Heaven' in shop windows, billboards, and on the t-shirts of pedestrians walking by. It clearly demonstrated how information floods into your brain without you paying any attention to it yet is available when a relevant question pops up. Addictive viewing for sure.

The power of priming and language

If I tell myself I am demotivated, guess what? I will be focusing on demotivation and not motivation. If there is no milk for my morning coffee and I am frustrated, then it doesn't take much for me to take that frustration into my work and project it into my first conversation.

We all know that we achieve goals through our ability, planning, hard work and motivation. But it is easy to derail your achievement with self-chatter. Thinking about how you can use priming to achieve motivation is a useful thing to learn and is limited only by your imagination. Who would have thought I would be talking to photos of my parents about my progress with this book? Sometimes role models help. Thinking of them can bust your negative thoughts and prime you into a better place. What would my best friend do in this situation? What would Richard Branson do? If your goal is about losing weight, a great way to start is to find some inspiring stories, maybe from someone you know or a celebrity or a cookery book writer like Tom Kerridge. There are so many ways to prime your way into a positive state. Read articles, books, watch programmes about role models, find someone who you can aspire to and want to work towards. Practice visualisation, affirmations and mental rehearsals as discussed in Chapter 1 on habit change. Reframe the lack of milk in the fridge as being a sign that today is about drinking more water. Remember, you are always in control of your responses to situations and you can choose to prime yourself in a positive way. In my little home office, I have a number of priming items that make me feel great, from beautiful stationery, multi-coloured sticky notes and pens, to huge golden metal wings on the wall that represent possibilities for me, to photographs of my husband and children, to my recent running medals, to client thank you notes. I also have my Alexa on standby for me to request any tune of my liking at any time.

A Stanford and Harvard University study found that subtle changes in language can change your voting behaviour.[28] The noun versus verb phrasing of survey items was varied to frame voting either as something you were, eg 'I am a voter', or as simply a behaviour, eg 'I am voting'. The personal-identity phrasing significantly increased interest in registering to vote and, in two state-wide elections in the US, increased voter turnout as assessed by official state records. The researchers conclude that these results provide evidence that we continually manage our self-concepts, seeking to assume or affirm valued personal identities.

The results demonstrate how language can be used to motivate socially relevant behaviour and to prime your responses to events.

Reach your goals through motivation

If you lack motivation, then it is unlikely you will take consistent action to achieve your goals, maybe starting but not completing, and certainly not getting to the stage of sustaining them. I can't count how many times I have heard clients say they want to lose weight for an important occasion, starting with gusto and massive enthusiasm, only to give up at a hurdle, a spontaneous barbecue, a party or a rainy day where they just can't be bothered to go for a jog. If you consider Sinek's golden circle and the 'must lose weight' brigade there is a particular category of weight losers who most often succeed with their goals – brides. And that is because they are super congruent in all aspects of their lives with their 'why'. Putting statistics aside and focusing on love, a wedding is a once in a lifetime event, with a single day of once in a lifetime photographs in a once in a lifetime incredible dress. The photos however are displayed for a full lifetime (and longer) so most brides are highly motivated to achieve that once in a lifetime body to slide into their dream gown. It all depends on how you frame it to get to the strongest why.

Motivation can be thought of as what initiates, guides and maintains goal-directed behaviour. It has momentum, direction, progression and propulsion. Motivation takes you from your desire for something through small steps of action, to bigger steps, through learning, navigating the journey, flexing, adapting then driving towards your goal achievement. This chapter has explained all about activation of the reward system, causing the release of dopamine which makes you feel good, want to keep going and repeat actions that made you feel this way. The whole process of working towards a goal can feel pleasurable when you have activated the brain and body's natural way of keeping intrinsically motivated.

But remember that motivation isn't a formula – as Mark discovered through our hours of coaching. Setting a goal that feels good of course isn't a guarantee of success. Nor is just having a strong why – even brides fail at weight loss before their weddings sometimes. As you move towards your goals, remember that success is a collection of many actions on your journey, the mental effort you apply, the preparation you take, the adjustments you make, the actions you perform every day – all glued together by your motivation.

Turn the heat up on your motivation by becoming consciously emotionally invested. Your emotions are like the directors of your personal motivational movie which rally you to keep going. In setting your goals, make them realistic, tangible and near, but keep the fire in your belly lit by engaging your why. It will be that fire that fuels you and sustains you on your journey. Explicitly setting your goals in line with your values and your why will help keep the fire blazing.

Set challenging and forward-looking but realistic goals. Exceeding expectations and seeing progress will help to stoke the fire while continually missing them will inevitably dampen the embers. See the goal, feel it and express it. Make it come as alive as possible in your brain.

Towards and away from language

An interesting study around health communications from Kent State University and the University of California, Santa Barbara offers food for thought.[29] The researchers found that receiving a message that was congruent with a long-term disposition of motivation for approach or avoidance resulted in stronger intentions and uptake of dental flossing. People with a strong approach (or towards) orientation are more responsive to cues of reward, where people with a predominant avoidance (or away from) orientation are more responsive to cues of threat or punishment. A person who veers to the approach disposition will respond better to the gain frame of flossing resulting in healthier gums while the loss frame of avoiding bad breath and gum disease will land more effectively for a person who has more of an avoidance temperament. For some people wanting to drop a dress size, keeping a beautiful dress on permanent display in a size smaller can be a fantastic motivator while for others, keeping their old jeans a size larger motivates them never to return to their previous overweight life.

As a coach, even if clients have a disposition to motivate away from something, I always work with them to find a 'towards' frame. This provides future direction and a positive vision of what can be achieved around the corner, a reframe of where they are now to where they are going. Large jeans or small dresses, the framing and language of goals to keep you motivated really matters. For Mark, his new role became a stepping stone for him to create and shape the global leadership career he desired and he had his 'why' words displayed as his screen saver on his devices as a reminder to keep him motivated.

Goal maintenance is all about laying down new neural pathways in establishing new habits and routines which is discussed in detail in Chapter 1. It's a great feeling when you can look back at the journey of working towards your goals and be proud of how you got there, but it's even better when you know this behaviour change you strived for is embedded and has become part of who you are now. And just in case I was in any doubt, I keep my 10k medals displayed in my office to remind me that at my grand old age, I can still achieve a dream.

CHAPTER SEVEN

Neuro Coach

'Because the brain is the body's captive audience, feelings are winners among equals.'
– Antonio Damasio, *Descartes Error. Emotion, Reason and the Human Brain*

What an exciting time it is to be a coach. The world continues to become more complex, brimming over with noise and stimulation flooding our senses in every waking moment, increasing the load and pressure on our brains. In the middle of this chaotic, inspiring, busy world sit our clients, full of challenges, possibilities and options. They seek coaches for a multitude of reasons to make sense of their place in the world and to work out how to stop being stuck. At the same time, we now have new insights, tools, wisdom and knowledge from neuroscience at our disposal to help our clients take a peek inside their brains.

So what?

So why does neuroscience matter? As coaches, we have successfully helped our clients for years, seeing them reach their goals, make valuable changes, become more self-aware, happier, more productive and successful. How does this new layer of knowledge add value to what is already working well? How does it make coaching an even better experience for both coach and client?

The added value of neuroscience to coaching

Neuroscience tells us that every brain is unique and therefore every client is unique. Intellectually we know this, but the science anchors that for us. Coaching is not generalised, it is bespoke, individual to each client giving rise to infinitely different conversations, insights and options. Learning about their unique brain can help clients to give themselves permission to be their authentic selves and to learn to appreciate their exclusive combination of strengths. That can be a liberating and empowering space to work from.

As Chapter 6 described, humans are innately curious beings and people want to understand what is going on inside their heads. When they understand that their brain is part of their stuck state and they can unpick and learn about the science behind it, then it can open them up, giving their problems a voice and a space for resolution. Having a physiological explanation can unlock new ideas and approaches for working through their challenges.

Coaching exposes the brain to learning and opens new pathways. Chapter 1 discussed forming new habits by activating and strengthening new neural circuits through repeated use. Engaging in coaching discussions can turn on neural pathways for self-reflection and discovery building a new habit for development, experimentation and learning.

Neuroscience introduces a layer of different, helpful language providing scope for deeper questioning or conversation. Examples include: 'If your amygdala calmed for a few moments what will happen?', 'What can you do now to give yourself a dopamine boost?', 'What is your nucleus accumbens tempting you with?', 'Focus on immediate feelings, let me be your cortex'. For some clients using metaphors paints a new landscape: 'The conductor of the orchestra [prefrontal cortex] is exhausted, no wonder decisions are tiring', 'What is the sabre tooth tiger in this situation?', 'Tell me why your brain is so full'.

Approaching their change work from a scientific rather than a behavioural perspective can help some clients to commit more deeply to the coaching process. They may discover an unexpected interest in the subject that can facilitate more invested engagement in the softer language of psychology and human behaviour interventions. An empirical, scientific approach can mean more trust in the 'soft stuff' and therefore energy to engage as there is a tangible aspect in the mix. Furthermore, they can then see a path from that 'soft stuff' to the 'harder stuff' of enhanced performance and results. Some clients only measure success by tangible outcomes, and neuroscience can provide the means to build faith in coaching as a critical medium to get there.

Providing the layer of neuroscience safely opens up a route to surface and discuss emotions. Understanding the evolutionary development of the brain, the importance of emotional, instinctive senses and the huge proportions of neurons in subcortical brain regions has added a dimension of relief and permission with a number of my clients. This is particularly useful if a client has been 'sent' to a coach, rather than proactively made the decision to seek some help. In one case, a client told me he felt he was released from tight shackles by putting a voice to what he had been battling with for so long. He said he finally realised it was OK to talk about what he was feeling rather than what he was thinking. It took a number of sessions to get there but the neuroscience unlocked his tightly padlocked emotional shackles.

A discussion about the brain helps a client to keep an open mind as it broadens the scope of the conversation, rather than narrowing it down. Remember that no part of the brain works in complete isolation. There are also multiple contexts in which parts of the brain play a role. Neuroscience therefore keeps the client in a questioning mode, continually creating the opportunity for further exploration and discovery.

As coaches, we are in the business of neuroplasticity. Understanding that the brain can re-wire, opening up new access routes, strengthening existing pathways and dampening non-serving ones can be liberating. Clients realise that not only is change possible but also that there are many potential options to move out of their stuck space.

For some clients, working with neuroscience adds substance and credibility to the coaching process. An empirical underpinning can legitimise the work we do with our clients, enabling them to give coaching the same respect as they would afford other business programmes.

Finally, neuroscience helps us as coaches to remain authentic, true to ourselves, affording us additional self-awareness of our own unique strengths and allowing us to surface blind spots. As there is so much yet to discover in this fascinating area, we will continue to question, learn, explore and grow long into the future.

Brain applications

I often find that clients have a smattering of knowledge about the brain and occasionally some are deeply knowledgeable and understand that using neuroscience can enhance their development. It is important, no matter how invested in the topic, that the client understands that there is rarely a case where the language of 'X researcher found Y so this means that Z happens'. Neuroscience triggers more questions rather than providing causal answers – which is why it is so enriching to the coaching experience.

Some clients will tell you what they know about 'the amygdala' or 'the reward centre' or 'the limbic system' and may believe that particular part of their brain is not working properly or is the cause of all their problems. That could be the case, however it may be helpful to be able to offer the client some additional perspective to play with. The list below is designed to help you access research pertaining to specific brain regions highlighted in the studies cited in the previous chapters. It draws together the key research around specific brain regions used in this book.

Remember that while a brain region may play a leading role in one movie scene, there is no blockbuster film without supporting actors, producers, directors, camera men, sound engineers, costume designers and so on. Plus that region will be required in multiple other scenes to make the movie a full story and meaningful for the viewer. It's all interconnected. Particular brain regions have relevance and roles in many different areas of presenting issues that clients bring to your coaching, therefore allowing you to offer a multitude of different research perspectives and insights to your conversations.

The list below clearly demonstrates this, showing how the different brain regions have applications across the six presenting issues you took a deep dive into in this book. And of course, there is a plethora of presenting issues you will have experience of working with, so I offer you just a starting point. The research studies are highlighted alphabetically by brain region within chapter by reference number (in brackets), so just scoot to the relevant point within the chapter for context or visit the notes section at the back of the book to see the formal book or research study title if you want to delve into more detail.

Amygdala: Chapter 2 (6, 7); Chapter 3 (1, 2, 3); Chapter 4 (23, 26, 32, 33); Chapter 5 (9, 31); Chapter 6 (24).

BDNF: Chapter 2 (2, 7).

Caudate: Chapter 6 (10, 22).

Cerebellum: Chapter 2 (12).

Cingulate cortex: Chapter 1 (8); Chapter 2 (10), Chapter 3 (1); Chapter 4 (11, 14, 23, 26, 31, 32); Chapter 5 (5); Chapter 6 (14).

Cortisol: Chapter 2 (2, 3, 4, 18).

Dopamine: Chapter 1 (5); Chapter 4 (24), Chapter 6 (8, 10, 15, 19, 21, 22, 25).

Dorsal raphe nucleus: Chapter 4 (24, 25).

Genes: Part 1 (1); Chapter 3 (7); Chapter 4 (30).

Habenula: Chapter 3 (6).

- Medial prefrontal cortex: Chapter 1 (8); Chapter 2 (6); Chapter 4 (31); Chapter 5 (33).

- Orbitofrontal cortex: Chapter 5 (9, 10, 11).

- Posterior temporal prefrontal cortex: Chapter 2 (12).

- Subgenual prefrontal cortex: Chapter 3 (5).

- Ventrolateral prefrontal cortex: Chapter 6 (20).

- Ventromedial prefrontal cortex: Chapter 1 (8); Chapter 5 (2, 33); Chapter 6 (11).

Putamen: Chapter 6 (19, 22).

Serotonin: Chapter 4 (25).

Striatum: Chapter 1 (5, 6, 7, 8, 10); Chapter 4 (8, 32); Chapter 5 (4, 5, 32, 33); Chapter 6 (10, 13, 19, 21).

Telomere: Chapter 2 (5, 18).

Temporo parietal junction: Chapter 4 (8, 31).

Thalamus: Chapter 5 (3).

Ventral tegmental area: Chapter 1 (5); Chapter 6 (21).

Conclusion

'I count him braver who overcomes his desires than him who conquers his enemies, for the hardest victory is over self.'
  – Aristotle

Is Aristotle right when it comes to coaching? Whether conquering being stuck turns out to be their hardest victory, a slow journey of discovery or based on a single eureka moment, I count all my clients as brave in wanting to create change. Enabling clients to reach a destination, to solve a problem, to give a voice to previously unarticulated turmoil, to become more self-aware, to make progress, however small, is a victorious salute to all coaches.

We are privileged in our profession to facilitate rich, enabling, sometimes life-changing conversations with our clients. And as described through the stories of

Peter, Jenny, John, David, Laura and Mark, neuroscience deepens and amplifies those conversations, providing additional perspective, empirical underpinning and a new way of helping clients understand what makes them tick and what they can do to when they are stuck.

I hope this book has provided food for thought, turned up the volume of your curiosity and added value to your work. Neuroscience will continue to yield important findings in the coming years in the quest to understand more about the 1.5-kilogram mass sitting in our skulls. For us coaches, this can only further enrich our lives and the experience we offer to our clients.

Appendix – MiND (MyBrain Indicator of Neurological Dominance)

Developed by MyBrain International, MiND is a tool that demonstrates the causal link between the preferences of a person and the neurology of their brain. MiND provides a unique 'neurometric' profile report, that is based on the latest neuroscientific research and which explains the physiological reasons behind everything you do, from the way you think, the types of goods you prefer to buy, the ways in which you communicate and interact with other people and even types of activities that energise you and those that drain you. It is therefore an invaluable resource for coaches, trainers, managers and HR professionals.

Using the MiND profiling tool offers you a snapshot of your neurological preferences and helps you to understand the 'why' question – 'why do I think this way', 'why do I react like this', 'why does this frustrate me', 'why am I different to the rest of my team'.

MiND stands apart from all other self-analytical tools in that as well as helping you understand what your personality, preferences, aptitudes or strengths are, it also explains how those traits arise as a result of the physical attributes of your brain. Becoming aware of this is hugely insightful as it can explain the reasons behind the differences that can exist between you and other people. As well as raising awareness, it inevitably raises more questions, encouraging you to explore deeper and further, so broadening out their insight.

MiND offers a unique approach by demonstrating that your preferences can vary in different circumstances. The MiND feedback report provides a breakdown of your overall profile into the preferences you exhibit in your personal life and those you show at work. Where differences exist, the MiND report can help you to understand this shift and may even enable you to utilise the differences to their advantage.

MyBrain International believes that offering you an accessible language around your preferences enables you to understand where you are operating at your most natural, truly in your flow, motivated, engaged and successful. There is nothing more interesting than understanding the basis of your behaviours and learning how your brain plays the most important role in who you are. Individuals who want to take a MiND profile questionnaire should get in touch via the web-site or email (details below).

Incorporating MiND into development workshops enables open discussion of colleagues' strengths and can result in significant improvements in areas such as employee motivation, self-confidence, openness as well as personal, team and organisational performance.

For coaches or trainers who are interested in using the MiND tool in their work, MyBrain International provides all the materials necessary as part of a comprehensive Practitioner training and support programme. Attending this enables sufficient knowledge to either run stand-alone activities for individual coaching clients or groups or to embed the tool within broader training and development programmes. Details are available on the company website or by email.

⊕ www.mybrain.co.uk
✉ enquiries@mybrain.co.uk

References

Part 1

1. Dennis, M.Y., Nuttle, X., Sudmant, P.H., Antonacci, F., Graves, T.A., Nefedov, M., Rosenfeld, J.A., Sajjadian, S., Malig, M., Kotkiewicz, H., Curry, C.J., Shafer, S., Shaffer, L.G., de Jong, P.J., Wilson, R.K., Eichler, E.E. (2012). Evolution of human-specific neural *SRGAP2* genes by incomplete segmental duplication. *Cell*, **149**(4), 912–922.

2. Herculano-Houzel, S. (2016). *The Human Advantage: A New Understanding of How Our Brains Became Remarkable*. Cambridge: MA. MIT Press.

3. Maguire, E.A., Gadian, D.G., Johnsrude, I.S., Good, C.D., Ashburner, J., Frackowiak, R.S.J., & Frith, C.D. (2000). Navigation-related structural change in the hippocampi of taxi drivers. *Proceedings of the National Academy of Sciences of the United States of America*, **97**(8), 4398–4403.

4. Goldberg E. (2009). *The New Executive Brain. Frontal Lobes In a Complex World*. New York. Oxford University Press.

5. Ibid.

6. Herculano-Houzel (n 2).

Chapter 1 – Old Habits Die Hard

1. Wood, W., Quinn, J.M., Kashy, D.A. (2002). Habits in everyday life: thought, emotion, and action. *Journal of Personality and Social Psychology.* **83** (6). 1281–1297.

2. Neal, D.T., Wood, W., Wu, M., Kurlander, D. (2011). The pull of the past: when do habits persist despite conflict with motives? *Personality and Social Psychology Bulletin,* **37** (11): 1428–1437.

3. Eagleman D. (2015), *The Brain: The Story of You.* Edinburgh. Canongate Books.

4. Eyal N. (2014). *Hooked: How To Build Habit-Forming Products.* London. Portfolio Penguin.

5. Newlin, D.B., Strubler, K.A. (2007). The habitual brain: an 'adapted habit' theory of substance use disorders. *Substance Use and Misuse, 42* (2–3): 503–526.

6. Atallah, H.E., McCool, A.D., Howe, M.W., Graybiel, A.M. (2014). Neurons in the ventral striatum exhibit cell-type specific representations of outcome during learning. *Neuron, 82*(5), 1145–1156.

7. McNamee, D., Liljeholm, M., Zika, O. O'Doherty, J.P. (2015). Characterizing the associative content of brain structures involved in habitual and goal-directed actions in humans: A multivariate fMRI study. *Journal of Neuroscience, 35* (9), 3764–3771.

8. Cascio, C.N. O'Donnell, M.B., Tinney, F.J., Lieberman, M.D., Taylor, S.E., Strecher, V.J., Falk, E.B. (2016). Self-affirmation activates brain systems associated with self-related processing and reward and is reinforced by future orientation. *Social Cognitive and Affective Neuroscience, 11* (4), 621–629.

9. Lally, P., van Jaarsveld, C.H.M., Potts, H.W.W., Wardle, J. (2010). How are habits formed: modelling habit formation

in the real world. *European Journal of Social Psychology, 40,* 998–1009.

10. Martiros, N., Burgess, A.A., Graybiel, A.M. (2018). Inversely active striatal projection neurons and interneurons selectively delimit useful behaviour sequences. *Current Biology,* 28 (4), 560–573.e5.

11. Duhigg, C. (2013). *The Power of Habit: Why We Do What We Do and How to Change.* London. Random House Books.

12. Schwartz, B. (2005). *The Paradox of Choice: Why More Is Less.* New York. New York. Harper Perennial.

13. Cirillo, F. (2018). *The Pomodoro Technique: The Life Changing Time Management System.* Ebury Publishing.

14. McGonigal, K. (2011). *The Willpower Instinct.* New York. New York. Avery Publishing.

Chapter 2 – The Typhoon of Stress

1. The Health and Safety Executive aims to reduce work-related death, injury and ill health. The 'About HSE' section of the website states: 'At the Health and Safety Executive, we believe everyone has the right to come home safe and well from their job. That's why our mission is to prevent work-related death, injury and ill health.' Labour Force Survey quoted, retrieved 6 August 2018 from www.hse.gov.uk/statistics/overall/hssh1617.pdf

2. Takuaz, D, Loya, A., Gersner, R., Haramati, S., Chen, A., Zangen, A. (2011). Resilience to chronic stress is mediated by hippocampal brain-derived neurotrophic factor. *Journal of Neuroscience,* **31**(12). 4475–4483.

3. Lara, V.P., Caramelli, P., Teixeira, A.L., Barbosa, M.T., Carmona, K.C., Carvalho, M.G., Fernandes, A.P., Gomes, K.B. (2013). High cortisol levels are associated with cognitive impairment no-dementia (CIND) and dementia. *Clinica Chimica Acta,* **423**, 18–22.

4. Herbert, J. (2013). Cortisol and depression: three questions for psychiatry. *Psychological Medicine*, **43**(3), 449–469.

5. Epel, E.S., Blackburn, E.H., Lin, J., Dhabhar, F.S., Adler, N.E., Morrow, J.D., Cawthon, R.M. (2004). Accelerated telomere shortening in response to life stress. *Proceedings of the National Academy of Sciences of the United States of America*, **101**(49), 17312–17315.

6. Kim, H., Yi, J.H., Choi, K., Hong, S., Shin, K.S., Kang, S.J. (2014). Regional differences in acute corticosterone-induced dendritic remodeling in the rat brain and their behavioral consequences. *BMC Neuroscience*, **15**, 65.

7. Ghosh, S. Laxmi, T.R. Chattarji, S. (2013). Functional connectivity from the amygdala to the hippocampus grows stronger after stress. *Journal of Neuroscience*, **33** (17), 7234–7244.

8. Charron, S. Koechlin, E. (2010). Divided representation of concurrent goals in the human frontal lobes. *Science*, **328** (5976) 360–363.

9. Janssen, C.P., Gould, S.J.J., Li, S.Y.W., Brumby, D.R., Cox, A.L. (2015). Integrating knowledge of multitasking and interruptions across different perspectives and research methods. *International Journal of Human-Computer Studies*, **79**, 1–5.

10. Loh, K.K., Kanai, R. (2014). Higher media multi-tasking activity is associated with smaller gray-matter density in the anterior cingulate cortex. *PLoS ONE* **9**(9), e106698.

11. Rubinstein, J.S., Meyer, D.E., Evans, J.E. (2001). Executive control of cognitive processes in task switching. *Journal of Experimental Psychology: Human Perception and Performance*, **27**(4), 763–797.

12. Lahnakoski, J.M., Jääskeläinen, I.P., Sams, M., Nummenmaa, L. (2017), Neural mechanisms for integrating consecutive and interleaved natural events. *Human Brain Mapping*, **38**, 3360–3376.

13. Covey, S., Merrill, A.R., Merrill, R.R. (1994). *First Things First: Coping With the Ever Increasing Demands of the Workplace.* London. Simon & Schuster UK.

14. Franklin Covey. 2017, August 24. Big Rocks (video file). Retrieved 6 August 2018 from https://youtube.com/watch?v= zV3gMTOEWt8

15. TED is a nonprofit devoted to spreading ideas, usually in the form of short, powerful talks (18 minutes or less). TED began in 1984 as a conference where Technology, Entertainment and Design converged, and today covers almost all topics – from science to business to global issues – in more than 100 languages. Retrieved 6 August 2018 from https://ted.com/about/our-organization

16. Pink, D.H. (2018). *When: The Scientific Secrets of Perfect Timing.* Edinburgh. Canongate Books.

17. Berchtold, N.C., Castello, N., Cotman, C.W. (2010). Exercise and time-dependent benefits to learning and memory. *Neuroscience,* **167**(3), 588–597.

18. Epel, E., Daubenmier, J., Moskowitz, J.T., Folkman, S. Blackburn, E. (2009). Can Meditation Slow Rate of Cellular Aging? Cognitive Stress, Mindfulness, and Telomeres. *Annals of the New York Academy of Sciences,* **1172**, 34–53.

19. Creswell, J.D., Taren, A.A., Lindsay E.K., Greco, C.M. Gianaros, P.J., Fairgrieve, A., Marsland, A.L., Brown, K.W., Way, B.M., Rosen, R.K., Ferris, J.L. (2016). Alterations in resting-state functional connectivity link mindfulness meditation with reduced interleukin-6: A randomized controlled trial. *Biological Psychiatry,* **80** (1), 53–61.

Chapter 3 – Negative Nightmares

1. Korb, A. (2015). *The Upward Spiral. Using Neuroscience to Reverse the Course of Depression One Small Change at a Time.* Oakland. California. New Harbinger Publications.

2. Goldberg E. (2009). *The New Executive Brain: Frontal Lobes In a Complex World*. New York. New York. Oxford University Press.

3. LeDoux, J. (1999). *The Emotional Brain*. New York. New York. Simon & Schuster.

4. Albert Ellis (1913–2007) was a 20th century psychologist who pioneered the development of Rational Emotive Behaviour Therapy, aimed to address the behaviours and belief systems of a client. This therapy is widely considered a precursor to the now popular Cognitive Behavioural Therapy.

5. Hamilton, J.P., Farmer, M., Fogelman, P. Gotlib, I.H. (2015). Depressive rumination, the default-mode network, and the dark matter of clinical neuroscience. *Biological Psychiatry,* **78** (4), 224–230.

6. Lawson, R.P., Seymour, B., Loh, E., Lutti, A., Dolan, R.J., Dayan, P., Weiskopf, N., Roiser, J.P. (2014). The habenula encodes negative motivational value associated with primary punishment in humans. *Proceedings of the National Academy of Sciences of the United States of America,* **111**(32), 11858–11836.

7. Todd, R.M., Müller, D.J., Lee, D.H., Robertson, A., Eaton, T., Freeman, N., Palombo, D.J., Levine, B., Anderson, A.K. (2013). Genes for emotion-enhanced remembering are linked to enhanced perceiving. *Psychological Science,* **24**(11), 2244–2253.

8. De Berker, A.O., Rutledge, R.B., Mathys, C., Marshall, L., Cross, G.F., Dolan, R.J., Bestmann, S. (2016). Computations of uncertainty mediate acute stress responses in humans. *Nature Communications,* **7**, 10996.

9. Hecht, D. (2013). The neural basis of optimism and pessimism. *Experimental Neurobiology,* **22**(3),173–199.

10. Daniel Goleman is an internationally known psychologist who lectures frequently to professional groups, business audiences, and on college campuses. As a science journalist Goleman reported on the brain and behavioural sciences

for *The New York Times* for many years. His 1995 book, *Emotional Intelligence* was on *The New York Times* bestseller list for a year-and-a-half, with more than 5,000,000 copies in print worldwide in forty languages and has been a best seller in many countries. Retrieved 6 August 2018 from www.danielgoleman.info/biography

11. David, S. (2016). *Emotional Agility: Get Unstuck, Embrace Change and Thrive In Work and Life*. London. Penguin Random House.

12. Taken from *The importance of feelings*, MIT Technology Review June 17, 2014. Neuroscientist Antonio Damasio explains how minds emerge from emotions and feelings in an interview with Jasin Pontin, Editor in Chief.

13. Wegner, D.M., Schneider, D.J., Carter, S.R., White, T.L. (1987). Paradoxical effects of thought suppression. *Journal of Personality and Social Psychology*, **53**(1), 5–13.

14. Porto, P.R., Oliveira, L., Mari, J., Volchan, E., Figueira, I., Ventura, P. (2009). Does cognitive behavioral therapy change the brain? A systematic review of neuroimaging in anxiety disorders. *Journal of Neuropsychiatry and Clinical Neurosciences*, **21** (2), 114–25.

15. De Berker, A.O., Rutledge, R.B., Mathys, C., Marshall, L., Cross, G.F., Dolan, R.J., Bestmann, S. (n 8).

Chapter 4 – Loneliness Hurts

1. From *Trapped in a Bubble* research from The Loneliness Action Group, led by the British Red Cross and Co-op partnership. The Loneliness Action Group was tasked by the Jo Cox Commission on Loneliness partners to secure a lasting legacy for the Commission's work.

2. Office for National Statistics (GB). (10 April 2018). *Loneliness. What characteristics and circumstances are associated with feeling lonely?* Analysis of characteristics and circumstances

associated with loneliness in England using the Community Life Survey, 2016 to 2017.

3. Holt-Lunstad, J., Smith, T.B., Layton J.B. (2010). Social relationships and mortality risk: A meta-analytic review. *PLoS Medicine* 7(7), e1000316.

4. Valtorta, N.K., Kanaan, M., Gilbody, S., Ronzi. S., Hanratty, B. (2016). Loneliness and social isolation as risk factors for coronary heart disease and stroke: Systematic review and meta-analysis of longitudinal observational studies, *Heart, 102* (13). Published Online First: 18 April 2016.

5. Holt-Lunstad, J., Smith, T.B., Baker, M., Harris, T., Stephenson, D., (2015). Loneliness and social isolation as risk factors for mortality; a meta-analytic review. *Perspectives on Psychological Science,* **10** (2) 227–237.

6. Cacioppo, J.T., Hawkley, L.C., (2009). Perceived social isolation and cognition. *Trends in Cognitive Sciences,* **13** (10) 447–454.

7. Cacioppo, S., Balogh, S., Cacioppo, J.T. (2015), Implicit attention to negative social, in contrast to nonsocial, words in the Stroop task differs between individuals high and low in loneliness: Evidence from event-related brain microstates. *Cortex,* **70**, 213–233.

8. Cacioppo, J.T., Norris, C.J., Decety, J., Monteleone, G., Nusbaum, H. (2009). In the eye of the beholder; Individual differences in perceived social isolation predict regional activation to social stimuli. *Journal of Cognitive Neuroscience,* **21** (1), 83–92.

9. Cacioppo, J.T., Patrick, W. (2008). *Loneliness: Human Nature and the Need For Social Connection.* New York. New York. W.W. Norton and Company, Inc.

10. Bowlby, J. (1997). *Attachment and loss, Volume 1.* London. Pimlico.

11. Eisenberger, N.I., Lieberman, M., Williams, K.D. (2003). Does rejection hurt? An fMRI study of social exclusion. *Science*, **302**, (5643), 290–292.

12. Kross, E., Berman, M.G., Mischel, W., Smith, E.E., Wager, T.D. (2011). Social rejection shares somatosensory representations with physical pain. *Proceedings of the National Academy of Sciences*, **108** (15) 6270–6275.

13. Taken from *Chronic loneliness is a modern day epidemic*, Fortune.com, June 22 2016. John Cacioppo, Director of the University of Chicago's Center for Cognitive and Social Neuroscience was interviewed by Laura Entis.

14. Onoda, K., Okamoto, Y., Nakashima, K., Nittono, H., Yoshimura, S., Yamawaki, S., Yamaguchi, S., Ura, M. (2010). Does low self-esteem enhance social pain? The relationship between trait self-esteem and anterior cingulate cortex activation induced by ostracism, *Social Cognitive and Affective Neuroscience*, **5** (4), 385–391.

15. Donovan, N.J., Okereke, O.I., Patrizia, V., Amariglio, R.E., Rentz, D.M., Marshall, G.A., Johnson, K.A., Sperling, R.A. (2016). Association of higher cortical amyloid burden with loneliness in cognitively normal older adults. *JAMA Psychiatry*, **73** (12), 1230–1237.

16. Holwerda, T.J., Deeg, D., Beekman, A.T.F., van Tilberg, T.G., Stek, M.L., Jonker, C., Schoevers, R.A. (2014). Feelings of loneliness, but not social isolation, predict dementia onset: results from the Amsterdam Study of the Elderly (AMSTEL*). Journal of Neurology, Neurosurgery and Psychiatry*, *85*, 135–142.

17. Taken from *The science of solitary confinement*, Smithsonian. com, 19 February 2014 written by Joseph Stromberg.

18. Social Psychologist Craig Haney, PhD, studies the use and impact of solitary confinement on inmates in super-maximum security, or 'supermax,' prisons. He is a social psychologist and a professor at the University of California, Santa Cruz.

19. Haney, C. (2003) Mental Health Issues in Long-Term Solitary and 'Supermax' Confinement. *Crime & Delinquency,* **49** (1), 124–156.

20. Bedrosian, T. A., Nelson, R. J. (2017). Timing of light exposure affects mood and brain circuits. *Translational Psychiatry,* **7** (1), e1017.

21. Ozcelik, H., Barsade, S. (2011). Work loneliness and employee performance. *Academy of Management Annual Meeting Proceedings,* (1), 1–6.

22. Eagleman D. (2015), *The Brain: The Story of You.* Edinburgh. Canongate Books.

23. Krill, A., Platek, S. (2009). In-group and out-group membership mediates anterior cingulate activation to social exclusion. *Frontiers in Evolutionary Neuroscience.* **1**, 1.

24. Matthews, G. A., Nieh, E. H., Vander Weele, C. M., Halbert, S. A., Pradhan, R. V., Yosafat, A. S., Glober, G. F., Izadmehr, E. M., Thomas, R. E., Lacy, G. D., Wildes, C. P., Ungless, M. A., Tye, K. M. (2016). Dorsal raphe dopamine neurons represent the experience of social isolation. *Cell,* **164** (4), 617–631.

25. Sargin, D., Oliver, D. K., & Lambe, E. K. (2016). Chronic social isolation reduces 5-HT neuronal activity via upregulated SK3 calcium-activated potassium channels. *eLife,* **5**, e21416.

26. Hsu, D. T., Sanford, B. J., Meyers, K. K., Love, T. M., Hazlett, K. E., Wang, H., Ni, L., Walker, S. J., Mickey, B. J., Korycinski, S. T., Koeppe, R. A., Crocker, J. K., Langenecker, S. A., Zubieta, J. K. (2013). Response of the μ-opioid system to social rejection and acceptance. *Molecular Psychiatry* **18,** (11), 1211–1217.

27. Caputo, A. (2015). The relationship between gratitude and loneliness: the potential benefits of gratitude for promoting social bonds. *Europe's Journal of Psychology,* **11**(2), 323–334.

28. Wood, A. M., Maltby, J., Gillett, R., Linley, P. A., Joseph, S. (2008). The role of gratitude in the development of social

support, stress, and depression: two longitudinal studies. *Journal of Research in Personality*, **42**, 854–871.

29. Emmons, R. A., & McCullough, M. E. (2003). Counting blessings versus burdens: an experimental investigation of gratitude and subjective well-being in daily life. *Journal of Personality and Social Psychology*, **84**(2), 377–389.

30. Algoe, S. B., Way, B. M. (2014). Evidence for a role of the oxytocin system, indexed by genetic variation in *CD38*, in the social bonding effects of expressed gratitude, *Social Cognitive and Affective Neuroscience*, **9** (12), 1855–1861.

31. Filkowski, M.M., Cochran, R.N., Haas, B.W. (2016). Altruistic behaviour mapping responses in the brain. *Neuroscience and Neuroeconomics*, **5**, 65–75.

32. Inagaki, T.K., Bryne Haltom, K.E., Suzuki, S., Jevtic, I., Hornstein, E., Bower, J.E., Eisenberger, N.I. (2016). The neurobiology of giving versus receiving support: The role of stress-related and social reward-related neural activity. *Psychosomatic Medicine* (78), **4**, 443–453.

33. Chang, S.W.C., Fagan, N.A., Toda, K., Utevsky, A.V., Pearson, J.M., Platt M.L. (2015). Neural mechanisms of social decision-making in the primate amygdala. *Proceedings of the National Academy of Sciences*, **112**, (52), 16012–16017.

34. Originally known for its public opinion polls since 1935, Gallup now focuses on providing analytics and management consulting to organisations globally.

35. 'Their research yielded Gallup's Q^{12} survey: the 12 questions that measure the most important elements of employee engagement. Gallup has studied survey results from more than 35 million employees around the world.'

36. Retrieved 6 August 2018 from https://q12.gallup.com/Public /en-us/Features

37. Gardner, W.L., Pickett, C.L., & Knowles, M. (2005). Social Snacking and Shielding: Using Social Symbols, Selves, and Surrogates in the Service of Belonging Needs. In K.D.

Williams, J.P. Forgas, & W. von Hippel (Eds), *The Social Outcast: Ostracism, Social Exclusion, Rejection, and Bullying* (pp. 227–242). New York: Psychology Press.

38. Brooks, H., Rushton, K., Walker, S., Lovell, K., Rogers, A. (2016). Ontological security and connectivity provided by pets: a study in the self-management of the everyday lives of people diagnosed with a long-term mental health condition. *BMC Psychiatry, 16*, (1), 409.

Chapter 5 – Decisions, Decisions, Decisions

1. Wansink, B., & Sobal, J. (2007). Mindless eating: the 200 daily food decisions we overlook. *Environment and Behavior, 39* (1), 106–123.

2. Rudorf, S., Hare, T., A. (2014). Interactions between dorsolateral and ventromedial prefrontal cortex underlie context-dependent stimulus valuation in goal-directed choice. *Journal of Neuroscience, 34*, (48) 15988–15996.

3. Ploran, E, J., Nelson, S.M., Velanova, K., Donaldson, D.I., Petersen, S.E., Wheeler, M.E. (2007). Evidence accumulation and the moment of recognition: Dissociating perceptual recognition processes using fMRI. *Journal of Neuroscience, 27, (44)*, 11912–11924.

4. Ito, M., Doya, K. (2015). Distinct neural representation in the dorsolateral, dorsomedial, and ventral parts of the striatum during fixed- and free-choice tasks. *The Journal of Neuroscience, 35, (8)*, 3499–3514.

5. Rudorf, S., Preuschoff, K., Weber, B. (2012). Neural correlates of anticipation risk reflect risk preferences. *Journal of Neuroscience, 32*, (47) 16683–16692.

6. Redish, A.D., Mizumori, S.J. (2014). Memory and decision-making. *Neurobiology of learning and memory, 117, 1–3.*

7. Pushkarskaya H., Liu X., Smithson M., Joseph J.E. (2010). Beyond risk and ambiguity: deciding under ignorance. *Cognitive Affective and Behavioral Neuroscience*, **10**, (3), 382–91.

8. Goleman D. (1996). *Emotional Intelligence, Why It Can Matter More Than IQ*. London. Bloomsbury Publishing.

9. Damasio A.R. 1994. *Descartes' Error: Emotion, Reason, and the Human Brain*. New York, New York. Putnam.

10. Bechara, A., Damasio A.R., Damasio H., Anderson S.W. (1994). Insensitivity to future consequences following damage to human prefrontal cortex. *Cognition*, **50**, 7–15.

11. Shiv, B., Loewenstein, G., Bechara, A., Damasio, H., Damasio, A.R. (2005). Investment behavior and the negative side of emotion. *Psychological Science*, **16**, (6), 435–439.

12. Baumeister R.F. and Tierney J. (2012). *Willpower: Rediscovering Our Greatest Strengths*. New York, New York. Penguin.

13. Muraven, M., Shmueli, D., Burkley, E. (2006). Conserving self-control strength. *Journal of Personality and Social Psychology*, **91**, (3), 524–537.

14. DeWall, C.N., Baumeister, R.F., Mead, N.L., Vohs, K.D. (2011). How leaders self-regulate their task performance: evidence that power promotes diligence, depletion, and disdain. *Journal of Personality and Social Psychology*, **100**, (1), 47–65.

15. Danziger, S., Levav, J., Avnaim-Pesso, L. (2011). Extraneous factors in judicial decisions. *Proceedings of the National Academy of Sciences*, **108**, (17), 6889–6892.

16. Ashby, N.J.S., Glöckner, A. Dickert, S. (2011). Conscious and unconscious thought in risky choice: Testing the capacity principle and the appropriate weighting principle of unconscious thought theory. *Frontiers in Psychology*, **2**, 261.

17. Kahneman D. (2011). *Thinking Fast and Slow*. London, England. Penguin Books.

18. Amos Tversky was a cognitive and mathematical psychologist, a student of cognitive science, a collaborator of Daniel Kahneman, and a figure in the discovery of systematic human cognitive bias and handling of risk. Much of his early work concerned the foundations of measurement. He was co-author of a three-volume treatise, *Foundations of Measurement* (recently reprinted). His early work with Kahneman focused on the psychology of prediction and probability judgement; later they worked together to develop prospect theory, which aims to explain irrational human economic choices and is considered one of the seminal works of behavioural economics.

19. Evans, J. St. B.T., Stanovich, K.E. (2013). Dual-process theories of higher cognition: Advancing the debate. *Perspectives on Psychological Science,* **8**, (3), 223–241.

20. Gigerenzer G., Gaissmaier W. (2011). Heuristic Decision Making. *Annual Review of Psychology* **62**, (1), 451–82.

21. Zeigarnik B., (1927). On finished and unfinished tasks. *Psychologische Forschung,* **9**.

22. Savitsky, K., Medvec, V.H., Gilovich, T. (1997). Remembering and regretting: The Zeigarnik effect and the cognitive availability of regrettable actions and inactions. *Personality and Social Psychology Bulletin,* **23**, (3), 248–257.

23. Jenkins, A.C., Hsu, M. (2017). Dissociable contributions of imagination and willpower to the malleability of human patience. *Psychological Science,* **28**, (7), 894–906.

24. Grossmann, I., Kross, E. (2014). Exploring Solomon's paradox: Self-distancing eliminates the self-other asymmetry in wise reasoning about close relationships in younger and older adults. *Psychological Science,* **25**, (8), 1571–1580.

25. Mousavi, S., Gigerenzer, G. (2014). Risk, uncertainty, and heuristics. *Journal of Business Research,* **67**, (8), 1671–1678.

26. Xie, L., Kang, H., Xu, Q., Chen, M.J., Liao, Y., Thiyagarajan, M., O'Donnell, J., Christensen, D.J., Nicholson, C., Iliff, J.J.,

Takano, T., Deane, R., Nedergaard, M. (2013). Sleep drives metabolite clearance from the adult brain. *Science*, **342**, (6156), 373–377.

27. Walker M. (2017). *Why We Sleep: The New Science of Sleep and Dreams*. London, England. Penguin Books.

28. Creswell, J.D., Bursley, J.K., Satpute, A.B. (2013). Neural reactivation links unconscious thought to decision-making performance. *Social Cognitive and Affective Neuroscience*, **8**, (8), 863–869.

29. Sir Roger Gilbert Bannister CH CBE was a British middle-distance athlete and neurologist who ran the first sub-4-minute mile. At the 1952 Olympics in Helsinki, Bannister set a British record in the 1500 metres and finished in fourth place. This achievement strengthened his resolve to become the first athlete to finish the mile run in under four minutes. He accomplished this feat on 6 May 1954 at Iffley Road track in Oxford.

30. Kahneman D. (n 17).

31. De Martino, B., Kumaran, D., Seymour, B., Dolan, R.J. (2006). Frames, biases, and rational decision-making in the human brain. *Science*, **313**, (5787), 684–687.

32. Hafenbrack, A.C., Kinias, Z., Barsade, S.G. (2014). Debiasing the mind through meditation: Mindfulness and the sunk-cost bias. *Psychological Science*, **25**, (2), 369–376.

33. Pushkarskaya, H., Smithson, M., Joseph, J.E., Corbly, C. Levy, I. (2015) Neural correlates of decision-making under ambiguity and conflict. *Frontiers in Behavioural. Neuroscience*. **9**, 325.

Chapter 6 – The Motivation Mission

1. Kibler, M.E. (2015). Prevent your star performers from losing passion for their work. *Harvard Business Review Blog*. January 14.

2. 'Michael E. Kibler is founder and CEO of Corporate Balance Concepts. Founded in 1993, CBC is a pioneer in Executive Coaching & Development, providing world-class expertise to industry leading companies and their senior executives around the globe. Through its leading-edge proprietary process and its extensive corporate expertise, CBC guides organizations and their executives to deliver breakthrough results in today's challenging, shifting environments.' Retrieved 6 August 2018 from https://corporatebalance.com/about-us/#history

3. The CIPD is the Chartered Institute of Professional Development and is the professional body for the HR and People Development professions. The CIPD host regular conferences and exhibitions for HR and L&D professionals.

4. Cary Cooper is co-founder of Robertson Cooper, alongside colleague Ivan Robertson. He is professor of Organisational Psychology at Manchester Business school and is recognised as a world leading expert on well-being.

5. From Managing presenteeism: a discussion paper (May 23, 2011). Centre for Mental Health. Retrieved 6 August 2018 from https://centreformentalhealth.org.uk/publications /managing-presenteeism

6. Figures from data collected by Robertson Cooper from a sample of nearly 40,000 employees in the UK. Retrieved 6 August 2018 from https://robertsoncooper.com/blog/what-is-presenteeism

7. Pink, D.H. (2010). *Drive: The Surprising Truth About What Motivates Us*. Edinburgh. Canongate Books.

8. Self-Determination Theory is a theory of motivation. It is concerned with supporting our natural or intrinsic tendencies to behave in effective and healthy ways. SDT has been researched and practised by a network of researchers around the world.

9. The theory was initially developed by Edward L. Deci and Richard M. Ryan and has been elaborated and refined by

scholars from many countries. Ryan, a clinical psychologist, is a Professor at the Institute for Positive Psychology and Education at the Australian Catholic University in Sydney, Australia. Deci is currently a professor at the University of Rochester in the Department of Clinical and Social Sciences in Psychology in Rochester, NY.

10. Retrieved 6 August 2018 from https://selfdeterminationtheory .org

11. Tricomi, E., DePasque, S (2016), *The Role of feedback in learning and motivation,* in Sung-il Kim, Johnmarshall Reeve, Mimi Bong (ed.) *Recent Developments in Neuroscience Research on Human Motivation (Advances in Motivation and Achievement, Volume 19)* Emerald Group Publishing Limited, pp. 175–202.

12. Ariely, D. (2016) *Payoff: The Hidden Logic That Shapes Our Motivations.* London. Simon & Schuster.

13. Murayama, K., Matsumoto, M., Izuma, K., Matsumoto, K. (2010). Neural basis of the undermining effect of monetary reward on intrinsic motivation. *Proceedings of the National Academy of Sciences, 107,* (49) 20911–20916.

14. Murayama, K., Matsumoto, M., Izuma, K., Sugiura, A., Ryan, R.M., Deci, E.L., Matsumoto, K. (2015). How self-determined choice facilitates performance: A key role of the ventromedial prefrontal cortex, *Cerebral Cortex, 25,* (5), 1241–1251.

15. Meng, M., Ma, Q. (2015). Live as we choose. The role of autonomy support in facilitating intrinsic motivation. *International Journal of Psychophysiology,* **98**, (3, 1), 441–447.

16. Albrecht, K., Abeler, J., Weber, B., Falk, A. (2014). The brain correlates of the effects of monetary and verbal rewards on intrinsic motivation. *Frontiers in neuroscience, 8,* 303.

17. Lee, W., Reeve J. (2013). Self-determined, but not non-self-determined, motivation predicts activations in the anterior insular cortex: an fMRI study of personal agency. *Social Cognitive and Affective Neuroscience, 8,* (5), 538–545.

18. Schultz, W. (2016). Dopamine reward prediction error coding. *Dialogues in Clinical Neuroscience*, **18**, (1), 23–32.

19. Simon Sinek has a simple but powerful model for inspirational leadership – starting with a golden circle and the question 'Why?'. His examples include Apple, Martin Luther King, and the Wright brothers… TED details about the talk – retrieved 6 August 2018 from https://ted.com/talks/simon_sinek_how_great_leaders_inspire_action

20. Sinek, S. (2011). *Start With Why: How Great Leaders Inspire Everyone to Take Action.* London, Penguin,

21. Hungarian-American psychologist Mihaly Csikszentmihalyi (born 29 September 1934) is a Hungarian-American psychologist. He conceived the psychological concept of flow, a highly focused mental state. He is the Distinguished Professor of Psychology and Management at Claremont Graduate University. He was the former head of the department of psychology at the University of Chicago and of the department of sociology and anthropology at Lake Forest College.

22. de Manzano, Ö., Cervenka, S., Jucaite, A., Hellenäs, O., Farde. L., Ullén, F. (2013). Individual differences in the proneness to have flow experiences are linked to dopamine D2-receptor availability in the dorsal striatum. *NeuroImage, 67.* 1–6

23. Yoshida, K., Sawamura, D., Inagaki, Y., Ogawa, K., Ikoma, K., Sakai, S. (2014). Brain activity during the flow experience: A functional near-infrared spectroscopy study. *Neuroscience Letters,* **573,** 30–34.

24. Gruber, M.J., Gelman, B.D., Ranganath, C. (2014). States of curiosity modulate hippocampus-dependent learning via the dopaminergic circuit. *Neuron,* **84,** (2),486–496.

25. Jepma, M., Verdonschot, R.G., van Steenbergen, H., Rombouts, S.A., Nieuwenhuis, S. (2012). Neural mechanisms underlying the induction and relief of perceptual curiosity. *Frontiers in Behavioural Neuroscience,* **6**, (5), 1–9.

26. Brown, S., Vaughan, C. (2009). *Play: How It Shapes the Brain, Opens the Imagination and Invigorates the Soul.* New York, New York, Penguin.

27. Pellis, S.M., Pellis, V.C., Bell, H.C. (2010). The function of play in the development of the social brain. *American Journal of Play,* **2** (3), 278–296.

28. Kerridge, T., *Tom Kerridge's Dopamine Diet: My Low Carb, Stay Happy Way To Lose Weight.* (2017) Bloomsbury Publishing, London, UK.

29. Since his television debut with *Derren Brown: Mind Control* in 2000, Brown has produced several other shows for the stage and television in both series and specials. His 2006 stage show *Something Wicked This Way Comes* and his 2012 show *Svengali* won him two Laurence Olivier Awards for Best Entertainment. He has also written books for magicians as well as the general public.

30. Channel 4 (2012). Derren Brown: Mind Control, Series 4: Episode 1. Retrieved 6 August 2018 from www.channel4.com /programmes/derren-brown-mind-control/videos/series-4/ animal-heaven/2917758818001.

31. Bryan, C.J., Walton, G.M., Rogers, T., Dweck. C.S. (2011). Motivating voter turnout by invoking the self. *Proceedings of the National Academy of Sciences,* **108**, (31), 12653–12656.

32. Sherman, D.K., Mann, T., Updegraff, J.A. (2006). Approach/ avoidance motivation, message framing, and health behavior: Understanding the congruency effect. *Motivation and Emotion,* **30**, (2), 165–169.

Acknowledgements

The irony isn't lost on me that several times I got stuck during the writing of this, my first book. At the risk of it sounding like an Oscar speech, there are many people who have supported me along the way who I would like to thank.

To my clients Peter, Jenny, John, David, Laura and Mark for giving me permission to share their stories. You know who you are and I hope I have represented you well.

Moray, my wonderful husband and partner in crime for always having my back and supporting me in whatever crazy adventure I choose to take.

My three children Gregor, Ellie, and Georgina for growing into such wonderful young adults, making me laugh every day and still wanting their uncool mum to be involved in their lives.

Alistair Schofield, my business partner and friend for his endless patience when I was working on this manuscript and for embracing our differences which allow us to work so well as a team.

My dearest friend Kim Varley for her fantastic illustrations and talent for interpreting content into images.

Ben Read at the G14 studio for teaching me that self-compassion isn't selfish.

Wei Hei Kipling and the members of the Ealing Eagles running club for showing me it is never too late to achieve a dream.

Shaa Wasmund, Matt Thomas, Andy Moss and all the Freedom Collection-ers for their indispensable knowledge, tips, techniques and encouragement to get this book from my brain onto the page.

Roger Waltham, Lucy McCarraher, Joe Gregory and Kate Latham at Rethink Press for their faith in my book concept, their endless patience and teaching me about the publishing world.

My labradors Gordon and Charlie who sat loyally at my feet keeping me company during hours and hours of research and writing.

To all the scientists out there whatever their specialism – neuroscience, biology, medicine, psychology, anthropology, sociology, economics – for their quest to discover, explore and continue to learn about the vast topic of human behaviour. Their research, passion, questions and hunger for knowledge is what fuels books like this and enables the curious to learn something new every day about how humans interact with this crazy, wonderful world we live in.

The Author

G ill McKay is a Human
Resources professional with
extensive experience gained from
a varied career focusing on helping individuals, teams, business
leaders and HR functions enhance
performance, build capability and
implement transformation initia-
tives. She has successful leader-

ship, management and field experience with a record
of achievement in coaching, learning and development,
HR consultancy, change management and communications, describing herself to be 'all about building
capability and confidence'.

Passionate about neuroscience, Gill is joint Managing Director of MyBrain International, the creators
of MiND, a unique neurometric profiling tool. The

company provides tools and materials to enable professional coaches and trainers to use the exciting, evolving subject of neuroscience in their work so they can gain better results with their clients. MiND enables people to become more effective by building self-awareness of their thinking and communications style through understanding their neurological preferences which strongly influence the way they are. With that knowledge, the company works with businesses to help individuals, teams and whole organisations become more confident, effective, energised, engaged and focused on their strengths and diversity.

In her work, Gill draws on her skills as a trained coach, mentor, facilitator and behavioural psychometric/neurometric assessor. Gill has long international experience and regularly works with diverse teams from other cultures. She has held both local and global roles in large blue chip organisations as well as smaller companies and has many years' experience in designing and delivering, coaching programmes, development workshops and conference speeches enjoying the opportunity for group discussion, networking, one-to-one work and ongoing learning and sharing.

🌐 www.mybrain.co.uk
✉ enquiries@mybrain.co.uk
f facebook.com/mybrain-international
in linkedin.com/in/gillmckay